THE NATIONAL ACADEMIES

Advisers to the Nation on Science, Engineering, and Medicine

The **National Academy of Sciences** is a private, nonprofit, self-perpetuating society of distinguished scholars engaged in scientific and engineering research, dedicated to the furtherance of science and technology and to their use for the general welfare. Upon the authority of the charter granted to it by the Congress in 1863, the Academy has a mandate that requires it to advise the federal government on scientific and technical matters. Dr. Ralph J. Cicerone is president of the National Academy of Sciences.

The **National Academy of Engineering** was established in 1964, under the charter of the National Academy of Sciences, as a parallel organization of outstanding engineers. It is autonomous in its administration and in the selection of its members, sharing with the National Academy of Sciences the responsibility for advising the federal government. The National Academy of Engineering also sponsors engineering programs aimed at meeting national needs, encourages education and research, and recognizes the superior achievements of engineers. Dr. Charles M. Vest is president of the National Academy of Engineering.

The **Institute of Medicine** was established in 1970 by the National Academy of Sciences to secure the services of eminent members of appropriate professions in the examination of policy matters pertaining to the health of the public. The Institute acts under the responsibility given to the National Academy of Sciences by its congressional charter to be an adviser to the federal government and, upon its own initiative, to identify issues of medical care, research, and education. Dr. Harvey V. Fineberg is president of the Institute of Medicine.

The **National Research Council** was organized by the National Academy of Sciences in 1916 to associate the broad community of science and technology with the Academy's purposes of furthering knowledge and advising the federal government. Functioning in accordance with general policies determined by the Academy, the Council has become the principal operating agency of both the National Academy of Sciences and the National Academy of Engineering in providing services to the government, the public, and the scientific and engineering communities. The Council is administered jointly by both Academies and the Institute of Medicine. Dr. Ralph J. Cicerone and Dr. Charles M. Vest are chair and vice chair, respectively, of the National Research Council.

www.national-academies.org

COMMITTEE ON THE FUTURE OF ARCTIC SEA ICE RESEARCH IN SUPPORT OF SEASONAL TO DECADAL PREDICTION

JACKIE RICHTER-MENGE (Co-Chair), Cold Regions Research and Engineering Laboratory
JOHN E. WALSH (Co-Chair), University of Alaska, Fairbanks
LAWSON W. BRIGHAM, University of Alaska, Fairbanks
JENNIFER A. FRANCIS, Rutgers, The State University of New Jersey
MARIKA HOLLAND, National Center for Atmospheric Research
SON V. NGHIEM, Jet Propulsion Laboratory
ROBERT RAYE, Shell Projects and Technology
REBECCA WOODGATE,* University of Washington

NRC Staff

KATIE THOMAS, Study Director
LAUREN BROWN, Research Associate
AMANDA PURCELL, Research Associate
ALEXANDRA JAHN, Christine Mirzayan Fellow
ELIZABETH FINKELMAN, Senior Program Assistant

*Member through June 2012

POLAR RESEARCH BOARD

JAMES W. C. WHITE (Chair), University of Colorado, Boulder
SRIDHAR ANANDAKRISHNAN, Pennsylvania State University, University Park
JULIE BRIGHAM-GRETTE, University of Massachusetts, Amherst
DAVID H. BROMWICH, Ohio State University, Columbus
JENNIFER A. FRANCIS, Rutgers University, New Brunswick, New Jersey
EILEEN E. HOFMANN, Old Dominion University, Norfolk, Virginia
BERNICE M. JOSEPH, University of Alaska, Fairbanks
AMY LAUREN LOVECRAFT, University of Alaska, Fairbanks
MOLLY E. MCCAMMON, Alaska Ocean Observing System, Anchorage
ELLEN S. MOSLEY-THOMPSON, Ohio State University, Columbus
GEORGE B. NEWTON, QinetiQ North America, Marstons Mills, Massachusetts
CARYN REA, ConocoPhillips, Anchorage, Alaska
VLADIMIR E. ROMANOVSKY, University of Alaska, Fairbanks
GAIUS R. SHAVER, Marine Biological Laboratory, Woods Hole, Massachusetts
ALLAN T. WEATHERWAX, Siena College, Loudonville, New York

Ex-Officio Members

JACQUELINE M. GREBMEIER, University of Maryland, Solomons
TERRY WILSON, Ohio State University, Columbus
DENEB KARENTZ, University of San Francisco

NRC Staff

CHRIS ELFRING, Board Director
LAURIE GELLER, Senior Program Officer
KATIE THOMAS, Associate Program Officer
LAUREN BROWN, Research Associate
ELIZABETH FINKELMAN, Senior Program Assistant

Preface

Arctic sea ice plays important roles in moderating global climate and influencing oceanic and atmospheric circulation. Reductions in Arctic sea ice thickness and extent have been observed over the past few decades and the trend of shrinking Arctic sea ice cover is expected to continue. While there are intrinsic limitations on Arctic sea ice predictability, some predictability appears to reside in the initial ice/ocean state and in the longer-term trend. However, our limited understanding of the coupled and complex interactions among Arctic sea ice, oceans, atmosphere, and land also hinders our ability to predict the rate and magnitude of future variations. In addition, although several efforts are under way to better understand the role of Arctic sea ice in the broader context of the Arctic climate system, and to forecast sea ice, there is also a need to better understand the role that sea ice plays beyond the polar region.

The National Research Council (NRC) Committee on the Future of Arctic Sea Ice Research in Support of Seasonal to Decadal Prediction was tasked to plan and conduct a workshop with the goal of fostering a dialogue between polar scientists, agency representatives, and stakeholders. The workshop focused on current major challenges in sea ice prediction. Workshop participants were asked to identify new methods, observations, and technologies that might advance seasonal to decadal sea ice predictive capabilities through improved understanding of the Arctic system (see Box 1.1 for full statement of task).

The need for sea ice predictions is driven by new challenges and opportunities created by a changing Arctic environment. Advancements are necessary to address the growing and increasingly urgent demands from a broad array of stakeholders, with concerns spanning various direct and indirect scientific, technological, and societal impacts such as the planning for new shipping ports, oil and gas exploration, and increased marine transportation, as well as widespread ecological changes.

The workshop was held May 9-10, 2012, at the University of Colorado in Boulder (Appendix B). Nearly 50 workshop participants represented a wide spectrum of expertise in the Arctic sea ice community. Key was the participation of a range of stakeholders including scientists together with end-user groups. This workshop organization helped focus the workshop discussions on scientific research needs in connection with end-user operations.

The output from the workshop served to inform the committee in the preparation of this report. It is expected that the report will be of interest to agencies with Arctic research programs (e.g., National Science Foundation, Office of Naval Research, National Aeronautics and Space Administration, U.S. Fish and Wildlife Service, and National Oceanic and Atmospheric Administration) as well as policy makers, nongovernmental organizations, and others concerned about climate change impacts in the North. It is also anticipated that agencies with polar operational and planning responsibilities, such as the U.S. Navy or the U.S. Coast Guard, will find the report of interest.

Many individuals contributed to this report. Particular thanks go to those who took time from their busy schedules to participate in the workshop. Whether as panel members, breakout group leaders, moderators, rapporteurs, or contributors, the participants readily demonstrated the advantages of an integrative approach in identifying and discussing complex issues. On behalf of the entire committee, we also want to express gratitude to those associated with the NRC staff who provided keen insights, able direction, and tremendous support to our endeavor. This includes Alexandra Jahn, a Christine Mirzayan Fellow, who prepared a background document that was distributed to all workshop participants, summarizing recent related work and activities. It also includes board director Chris Elfring, research associate Lauren Brown, project assistants Amanda Purcell and Elizabeth Finkelman, and especially our exceptional study director Katie Thomas. Last but certainly not least is a word of thanks to the dedicated committee members, who volunteered countless hours to this effort and enthusiastically contributed their expertise to the organization and implementation of the workshop and preparation of this report.

Jackie Richter-Menge and John Walsh,
Co-Chairs
Committee on the Future of Arctic Sea Ice Research in Support of Seasonal to decadal Prediction

Acknowledgments

This report was reviewed in draft form by individuals chosen for their diverse perspectives and technical expertise in accordance with the procedures approved by the National Research Council (NRC) Report Review Committee. The purpose of this independent review is to provide candid and critical comments that will assist the NRC in making its published report as sound as possible, and to ensure that the report meets NRC institutional standards for objectivity, evidence, and responsiveness to the study charge. The review comments and draft manuscript remain confidential to protect the integrity of the deliberative process. We thank the following individuals for their review of this report:

LCDR Ken Boda, U.S. Coast Guard
Jennifer Kay, National Center for Atmospheric Research
Brendan Kelly, Office of Science and Technology Policy
Humfrey Melling, Fisheries and Oceans Canada
Don Perovich, Cold Regions Research and Engineering Laboratory
Caryn Rea, ConocoPhillips
Mike Steele, University of Washington

Although these reviewers provided many constructive comments and suggestions, they did not see the final draft of the report before its release. The review of this report was overseen by Robin Bell, Lamont-Doherty Earth Observatory of Columbia University. Appointed by the NRC, she was responsible for making certain that an independent examination of this report was carried out in accordance with institutional procedures and that all review comments were carefully considered. Responsibility for the final content of this report rests entirely with the authoring committee and the institution.

Contents

Recent changes in Arctic sea ice cover are of interest to a wide variety of stakeholders, and there is an increasing demand for improved sea ice predictability. This image shows an example of pancake ice off the coast of Greenland. Image courtesy of Andy Mahoney.

Summary

An understanding of Arctic sea ice is important because the Arctic plays a role in influencing not only the global climate system, but also the global economic system through changes in marine access and natural resource development (Box S.1). Recent dramatic changes in the thickness and extent of the Arctic sea ice cover, which can be linked to the warming climate, are well documented. These changes affect a growing community of diverse stakeholders, including local populations (e.g., indigenous populations), natural resource industries, fishing communities, commercial shippers, marine tourism operators, national security organizations, regulatory agencies, and the scientific research community. Accompanying this growing interest is an urgent demand to increase the pace and scope of the advancements in predictive capabilities. Added pressure comes with the reality of the limited resources (e.g., funding and infrastructure) available to enable continued improvement of Arctic sea ice prediction and the likelihood that, in the face of continued warming, the Arctic will remain a dynamic environment well into the 21st century.

As tasked by the National Aeronautics and Space Administration, the Office of Naval Research, and the intelligence community, the committee convened a

BOX S.1
KEY SCIENCE QUESTIONS

- What are the implications of the recent dramatic shifts in the Arctic from predominantly multiyear ice to first-year ice, and how will the associated complexities of this **regime shift** affect sea ice variability and predictability?
- In a rapidly changing Arctic regime, how will **forcings and couplings** between the various components of the ocean, atmosphere, cryosphere, seafloor, and land systems modify or influence the processes governing the characteristics of the sea ice cover?
- What are the impacts of **extreme events and feedback mechanisms** on Arctic sea ice evolution and our ability to predict it?
- How will **changing Arctic sea ice characteristics and dynamics** affect stakeholders on a variety of timescales, including prediction requirements?

workshop with the goal of exploring current major challenges in sea ice prediction and identifying new methods, observations, and technologies that might advance seasonal to decadal sea ice predictive capabilities through improved understanding of the Arctic system. The content of this report is largely informed by the discussions held during the workshop and is augmented by the committee's deliberations.

A key theme resonating throughout the report is the importance of a coordinated and integrative approach to advance sea ice prediction. In fact, fundamental to the success of the workshop was a purposeful approach taken to foster a dialogue between polar researchers, agency representatives, and end users. The committee concludes that there is a need for this dialogue to be sustained well beyond the confines of the workshop format. A committed and deliberately integrative approach, founded on a sustained and coordinated conversation among the user, modeling, and observation communities, is necessary to:

- Identify and address key gaps in our fundamental understanding of the Arctic environment and its connection to the global climate system;
- Balance high-priority stakeholder needs against realistic predictive capabilities;
- Foster coordinated support of this work within the private and public sectors;
- Provide guidance in allocation of resources to support the most promising avenues in addressing the most pressing needs.

This deliberately integrative approach would not only help address the challenges identified in Chapter 2, but is also necessary to effectively implement many of the strategies laid out in Chapter 3.

In this spirit, there are several key overarching challenges, not unique to the topic of sea ice prediction, that hinder advancements in our predictive capabilities:

- Treating the Arctic sea ice cover not in isolation, but as an integral part of the complex Arctic system which, in turn, is an integral element of the global system;
- Understanding how the recent regime shift in the Arctic sea ice cover from predominantly multiyear to first-year ice affects the processes governing the atmosphere-sea ice-ocean system, the power of statistical prediction methods, the validity of current numerical models and their parameterizations, and observational requirements, including instrument design; and
- Clearly defining the needs of the growing number of stakeholders, many with additional and more sophisticated requirements, and balancing these needs against realistic predictive capabilities.

The detailed needs of the diverse stakeholder community are reflected in an equally diverse set of temporal and spatial requirements. Likewise, many of the needs and challenges associated with sea ice prediction depend on the timescales of interest. At shorter timescales (seasonal to interannual), predictive capability is thought to reside primarily in an adequate knowledge of the initial ice-ocean state, although admittedly little information exists on what constitutes an "adequate knowledge." Challenges on the seasonal timescales include:

- Understanding the relative strengths and weaknesses of the different existing approaches used to generate seasonal ice forecasts (statistical algorithms, coupled ice-ocean models driven by prescribed atmospheric forcing, and coupled atmosphere-ocean-ice models);
- Establishing specific key observational data requirements necessary for defining the initial ice-ocean state for seasonal sea ice predictions; and
- Providing access to observational data at fast turnaround times.

At longer (decadal and greater) timescales, the role of trends in external forcings (e.g., increasing greenhouse gases) and of factors that control the forcings is likely to provide some predictive potential because they account for increasingly large fractions of the change from present sea ice conditions. A critical point of uncertainty remains regarding the timescale at which a transition occurs between these two regimes, and there is likely to be an intermediate timescale for which the potential predictability is low. The primary challenge at these longer timescales is to improve the ability to simulate realistic forcings by the atmosphere and ocean using coupled climate models at decadal timescales, and to identify the model variable and/or processes that contribute to unrealistic simulations.

In light of these challenges and while recognizing that there are limitations in current modeling and observational techniques, the committee offers possible strategies to significantly enhance our understanding and predictions of the Arctic sea ice cover over seasonal to decadal timescales:

- A systematic evaluation of the existing seasonal prediction capabilities to establish baseline expectations for predictive power and to set the stage for advances in predictive capability;
- A highly coordinated and integrated process-based study of seasonal sea ice focused on understanding the impact of the increasing predominance of younger, first-year ice on sea ice predictions and offering an opportunity to identify, develop, and test instruments and observational platforms;
- Inform research investments related to observational needs (e.g., observation types, locations, and coverage) in support of sea ice modeling and prediction by conducting an organized set of model sensitivity studies.
- Enhance the capabilities of numerical models through a coordinated experiment with multiple models to (a)

identify which variables and processes are critical to simulating a realistic ice cover, (b) determine the sources of climate model drift, and (c) guide decisions regarding high-priority model development needs and the expansion of models to include additional capabilities and variables of interest to stakeholders; and
- Create a centralized information hub that facilitates the timely access to observational and modeling results and encourages sustained communication among stakeholders.

These strategies are offered as guidance toward facilitating a transformative change in (1) our understanding of Arctic sea ice predictability on seasonal to decadal timescales and (2) our collective ability to realize and effectively communicate useful predictive power. The rate of advancement in sea ice predictions will likely be determined by the extent to which the broad user, modeling, and observational communities can achieve a sustained, integrative approach to refining and implementing these and other strategies.

1
Introduction

The Arctic is a region of increasing strategic and economic importance. Its influence spans a diverse array of stakeholders across international boundaries, including local populations (e.g., indigenous populations), natural resource industries, fishing communities, commercial shippers, marine tourism operators, national security organizations, regulatory agencies, and the scientific research community (e.g., Arctic Council, 2009). The Arctic also plays a number of roles in moderating global climate by influencing the planetary heat budget and interacting with the oceanic and atmospheric circulation systems as well as the terrestrial environment (Guemas and Salas-Melia, 2008; Lawrence et al., 2008; Deser et al., 2010; AMAP, 2011; Francis and Vavrus, 2012; Jakobsson et al., 2012; Koenigk et al., 2012; Maslowski et al., 2012; Nghiem et al., 2012).

The extent and thickness of Arctic sea ice has recently undergone extraordinary decline (Figure 1.1) that can be linked to climate changes (e.g., IPCC, 2007; Min et al., 2008; Allison et al., 2009; NRC, 2010; Kay et al., 2011; Notz and Marotzke, 2012;). The last six summers (2007-2012) have experienced the six lowest sea ice extent minima over three decades of satellite record, and the past decade (2003-2012) has exhibited 9 of the 10 lowest minima (updated from Perovich et al., 2011). The reduction of summer sea ice extent has been greatest in the Beaufort and Chukchi seas offshore of Alaska and in the Kara and Laptev Seas north of Russia. These are regions of particular interest to stakeholders concerned with marine access to subsistence activities, shore infrastructure, marine transportation, and natural resource developments. The winter sea ice extent has also shown a downward, though less striking, trend. More notable than winter sea ice extent is the change in composition of the winter sea ice cover, associated with the reduced summer sea ice. The winter sea ice cover now includes a significantly higher percentage of thin, seasonal (i.e., first-year) ice. The fraction of the first-year sea ice in the Arctic Ocean in March increased from 38 percent in the early 1980s to 64 percent in 2010 (Stroeve et al., 2011).

Closely associated with the persistent changes in the characteristics of the ice cover are other observed changes throughout the Arctic system. For instance, in regions of sea ice loss the upper ocean is warmer and fresher (e.g., Jackson et al., 2010; Steele et al., 2011). As a result of increased open water area, biological productivity at the base of

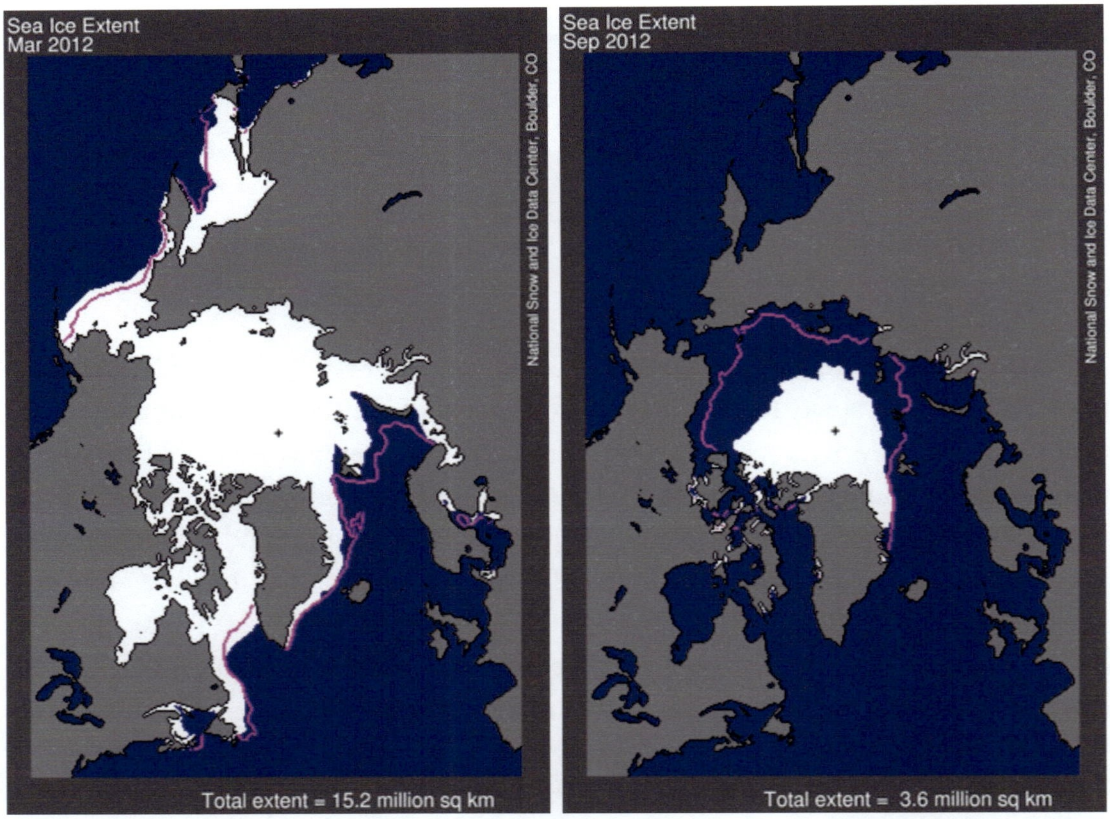

FIGURE 1.1 Arctic sea ice extent has recently undergone extraordinary decline: March 2012 (left) and September 2012 (right). The two periods that define the annual sea ice extent cycle are March, at the end of winter, when the ice is at its maximum extent, and September, at the end of summer, when the ice reaches its annual minimum extent. The purple line indicates the median maximum and minimum ice extents in the given month for the period 1979-2000. Compared with the 1979-2000 average, the September 2012 minimum was 49 percent smaller. SOURCE: Updated from Fetterer et al. (2002), Sea Ice Index, National Snow and Ice Data Center.

the marine food chain has increased (e.g., Arrigo and van Dijken, 2011) and sea ice-dependent marine mammals continue to lose habitat (e.g., Thomas and Laidre, 2011). Increases in the greenness of tundra vegetation and permafrost temperatures are linked to warmer land temperatures in coastal regions, often adjacent to the areas of greatest sea ice retreat (e.g., Bhatt et al., 2010).

Given the expectation for a continued increase in the global temperature throughout the 21st century (e.g., Meehl et al., 2007), the declining trends in Arctic sea ice extent and multiyear ice composition are expected to continue. Associated changes will likely result in greater marine access to the Arctic (e.g., for commercial shipping and offshore natural resource development) and increased coastal erosion, as well as a range of local, regional, and hemispheric changes in the climate and ecological systems.

The current and projected conditions and activity levels in the Arctic call for an improvement in our ability to predict characteristics of the sea ice cover. At the foundation of this challenge lies a set of specific questions related to sea ice prediction common among the broad stakeholder community: (1) Where is the ice at any given time (extent and concentration)? (2) What is it like (thickness distribution, multiyear ice versus first-year ice, packed or loose)? (3) What is its movement? (4) How and why has it changed? A major complicating factor in devising a strategy to more effectively address these questions is associated with the detailed and differing needs of the stakeholders, which shapes temporal and spatial requirements. For instance, while decadal projections of pan-Arctic ice extent and composition may serve the needs of decision makers who determine classification of endangered species or of planners who decide whether to build an ice-worthy ship, the long-range predictions will be less useful to marine operators whose main concern is the position of the ice edge. Springtime whaling and walrus hunting by coastal communities is dependent on over-ice transportation, requiring seasonal projections for the stability of the ice cover, its thickness and roughness, and the presence of open water close to shore. Detailed local information of this kind is not presently captured by monthly-to-seasonal ice forecasts and is only beginning to be targeted by experimental sea ice forecasts (e.g., Sea Ice Walrus Outlook[1]).

Many of the needs and challenges associated with sea ice prediction depend on the timescales of interest. At shorter timescales (seasonal to interannual), internal variability will severely limit sea ice predictability. For example, summertime wind patterns are known to impact summer and autumn ice extent, yet weather patterns are essentially unpredictable beyond about 2 weeks. Additional atmospheric observations are not going to change this fundamental prediction barrier. Rather, predictive capability for sea ice on these timescales is thought to reside primarily in an adequate knowledge of the initial ice-ocean state, although admittedly little information exists on what constitutes an "adequate knowledge." At longer (interannual to decadal and beyond) timescales, the role of trends in forcing (e.g., rising concentrations of greenhouse gases, changes in ocean mixing, increases in river discharge) is likely to provide some predictive potential, as it accounts for increasingly larger fractions of the change from present sea ice conditions.

A critical gap of uncertainty remains regarding the timescale between these two regimes, in which the potential predictability is low. Blanchard-Wrigglesworth et al. (2011) addressed these various prediction timescales in a numerical modeling study and suggested that a time frame of minimal predictability (for total Arctic Basin ice area) occurs from about 2 to 4r years. Although this study represents a state-of-the-art assessment of sea ice predictability, the results are based on one model (Community Climate System Model [CCSM4]) and are contingent on the ability of that model to capture the spectrum of oceanic, atmospheric, and terrestrial variability that affects sea ice. Hence the robustness of this

[1] http://www.arcus.org/search/siwo

result across climate models is uncertain, and the timescales are likely to vary based on the sea ice property, simulated atmospheric and oceanic forcings, and region of interest. Nevertheless, the existence of a minimum in predictability at interannual timescales of several years is plausible and likely. For this reason, the report focuses on seasonal and decadal predictability, with the understanding that improvements in predictions over these two timescales may eventually extend into the interannual timescale of relatively low predictability.

The particular challenges confronting the prediction of the character and behavior of sea ice in the seasonal time frame are compounded by the increasingly urgent need for this information by a variety of stakeholders. As noted above, coastal regions and the marginal ice zone are an important focus of these needs because sea ice affects access to shore infrastructure, marine transportation, resource extraction, and fishing activities. At these timescales stakeholders require rapid access to the information, and errors in that information tend to have immediate and often serious consequences. Furthermore, the regions and sea ice properties of interest will likely continue to experience fundamental change accompanying long-term variations in climate.

On decadal timescales, recent work has highlighted the considerable variability of Arctic sea ice. Kay et al. (2011), for example, found that different ensemble simulations from a single climate model subjected to the same changing external forcing (e.g., greenhouse gases and solar variability) exhibited a considerable spread in the simulated late 20th century Arctic ice loss. Moreover, some simulations for the 21st century revealed decadal-scale periods of gains in ice extent when internal variability counteracted greenhouse gas-forced trends. However, the chaotic internal variability can also reinforce the trend in ice loss, leading to instances of very rapid sea ice retreat (e.g., Holland et al., 2006). This large internal variability provides an inherent limit to predictability at these timescales, and as such, any decadal-scale predictions are necessarily probabilistic in nature. To date, however, little work has been done to assess the inherent limits on decadal predictability for different sea ice properties, at different times of year, and in different regions.

Some global coupled climate models are able to realistically simulate the past behavior of Arctic sea ice (e.g., Jahn et al., 2012). Figure 1.2 compares a single realization by the Community Climate System Model Version 3 (CCSM3) with observations of the actual ice cover, demonstrating the model's success in capturing not only the decadal-scale pace of ice loss, but also realistic interannual variability. Of particular note is the ability of the modeled ice extent to undergo a decade of recovery within the inexorable downward trend. Climate models that were discussed in the 4th Intergovernmental Panel on Climate Change (IPCC) assessment were compiled under the Coupled Model Intercomparison Project, Phase 3 (circa 2006). Simulations with newer models (circa 2011) that are informing the 5th IPCC assessment have just recently become available in CMIP5. As the models have progressed over this time,

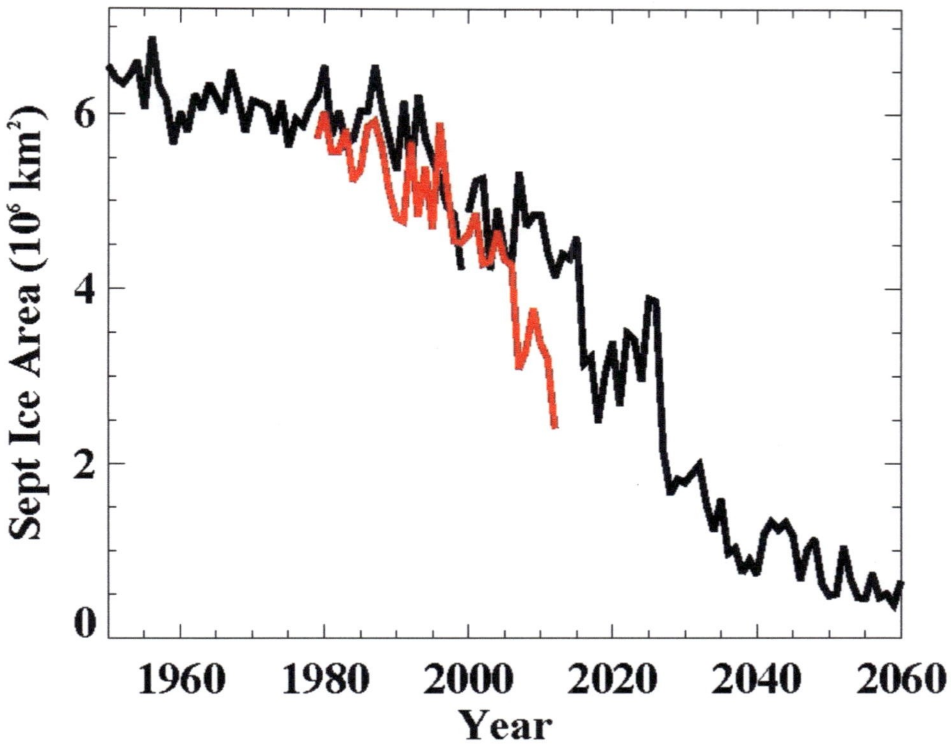

FIGURE 1.2 The 20th to 21st century September ice area in the Northern Hemisphere from a single CCSM3 ensemble member (black line) compared with the observed time series of September ice area from Fetterer et al. (2009, red line). It demonstrates the model's success in capturing the decadal-scale pace of ice loss and realistic interannual variability SOURCE: Adapted from Holland et al. (2011).

the realism of simulated sea ice area in response to anthropogenic and natural forcing has improved in comparisons with satellite observations available since 1979 (Massonnet et al., 2012; Stroeve et al., 2012).

Many climate models still simulate an Arctic ice pack at odds with observations. Figure 1.3 displays dozens of model runs from CMIP5 under two different future forcing scenarios termed "representative concentration pathways" (RCPs; Moss et al., 2010) including RCP4.5 and RCP8.5, which reach 4.5 W·m^{-2} (watts/square meter) and 8.5 W·m^{-2} radiative forcing by 2100, respectively. Most simulations capture a long-term reduction in ice extent that is driven fundamentally by external forcing—primarily increasing concentrations of greenhouse gases. The spread among the runs, however, has not decreased appreciably since the IPCC-4 generation, suggesting that basic challenges remain to be overcome. Because so many variables affect the ice evolution, identification of the causes

BOX 1.1
Statement of Task[a]

An ad hoc committee will plan and conduct a public workshop that outlines the current state of Arctic sea ice research, discusses knowledge gaps, and identifies emerging or important new science questions for the coming decade. Through invited presentations and discussion, participants will examine current observations and modeling efforts of sea ice, and identify (but not prioritize) areas of research and technology advances needed to better understand current and future changes. The committee will examine Arctic sea ice prediction, with a particular emphasis on seasonal to decadal timescales. The workshop will be designed to bring together polar scientists and agency representatives to explore whether there are new capabilities and infrastructure available to study sea ice in different ways that might shed new light on emerging research questions. This information may provide the context for future planning and policy development for sea ice research activities. The outcome of this activity will be a consensus report of the committee that builds on workshop presentations and discussions to provide conclusions on the following topics:

- What key scientific questions remain and how can we improve our understanding of the coupling between oceans, atmosphere, and sea ice (e.g., on what processes should observations be focused)?
- What systems of monitoring and observations are needed to better understand and predict the connection between changes in Arctic sea ice and its impacts on climate?
- What aspects of coupled sea ice models do we understand the best and in what ways can models better utilize existing observations and monitoring of sea ice to enhance our understanding of processes and future changes, and improve sea ice prediction?

[a] This report is sponsored by NASA, ONR, and the intelligence community.

of discrepancies in a particular model is often elusive, and sometimes improvements to one process reveal problems in other processes that were originally masked because of compensating errors.

As discussed previously, Arctic sea ice prediction has inherent limitations due to the chaotic nature of the climate system that may severely limit the possible predictive power. However, these limitations are poorly understood, especially across the full range of timescales and variables of interest to stakeholders. Our ability to realize the inherent predictability that does exist is further hindered by a limited understanding of the coupled and complex interactions between Arctic sea ice, oceans, and the atmosphere. Advances in understanding and seasonal to decadal predictive capabilities require enhancements of our theoretical,

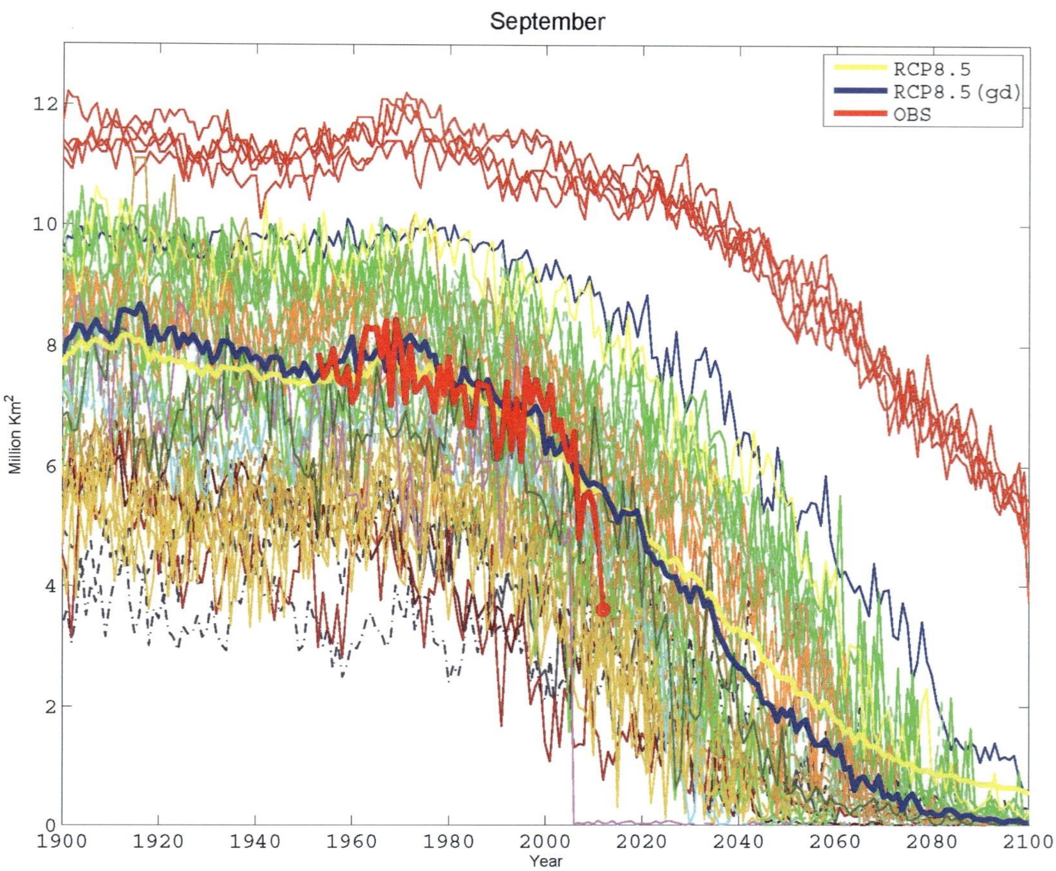

FIGURE 1.3 Many climate models still simulate an Arctic ice pack that differs from observations. This figure shows simulations of September sea ice extent from 1900 to 2100 by 23 global climate models participating in the 5th IPCC assessment. Each thin colored line represents one ensemble member. The thicker solid red line depicts observed ice extent, based on the Hadley estimate; except for September 2012 which is from NSIDC. The wide spread among the runs has not decreased appreciably since the 4th IPCC assessment, and because of the complexities associated with sea ice evolution, there are challenges that remain to be overcome. SOURCE: Wang and Overland, 2012. Reproduced/modified by permission of American Geophysical Union.

observing, and modeling capabilities. Evidence of the high level of concern about these limitations and the challenges involved in addressing them is demonstrated by the many recent studies that have focused on identifying key questions and recommendations related to Arctic sea ice predictability (see Appendix A for a summary of recent efforts). This report seeks to build on these efforts, with a specific emphasis on improved integration between the diverse community of stakeholders with a keen interest in and significant requirement for improved sea ice

BOX 1.2
Terms

Couplings—Two-way interaction between different subsystems (e.g., atmosphere, cryosphere, hydrosphere, etc.).

Decadal scale (4 to 30 years)—The term "projection" is commonly used when referring to this timescale.

Feedbacks—A sequence of interactions that determines the response of a system to an initial disturbance.

First-year ice—Floating ice of no more than 1 year's growth developing from young ice; thickness from 0.3 to 2 m (1 to 6.6 ft) where level; ridges of much thicker ice, to 30 m (98 ft), can form where floes are fractured by pressure, and these are rough and sharply angular.

Forcings—External data input into models that drive variability and change (e.g., solar radiation, greenhouse gas concentrations, volcanic aerosols).

Interannual scale (1-4 years) - The term "prediction" is generally used, although predictions for this timescale presently show little skill relative to climatology or persistence of trend.

Internal variability—Interactions within the climate system, as opposed to those forced externally (e.g., by changing greenhouse gas concentrations, solar variability, volcanic aerosols). Examples of internal variability include El Niño, the Arctic Oscillation, and the enhancing or dampening effects of feedbacks.

predictions on the seasonal to decadal timescale.

This report was developed from insights and information gained during a workshop. The goal of the workshop was to foster a dialogue among stakeholders (e.g., Arctic indigenous residents, polar scientists, agency representatives, and commercial interests) to explore current major challenges in sea ice prediction, and to identify new methods, observations, and technologies that might advance seasonal to decadal sea ice predictive capabilities through improved understanding of the Arctic system. The most prominent theme to emerge from the workshop was the idea of a committed and deliberately integrative approach to Arctic sea ice prediction that would require a

Numerical models—Numerical models solve systems of equations describing the fundamental physics, fluid motion, and thermodynamics of an Earth system component. These models can include single Earth system components (i.e., sea ice, ocean, land, or atmosphere) or can include multiple components that are coupled through the exchange of heat, water, and momentum (i.e., ice-ocean models, global climate models). Biogeochemistry, chemistry, and other aspects can also be incorporated through the inclusion of additional coupled equations or parameterizations.

Marginal ice zone—A band of pack ice 100 to 200 km (62 to 124 mi.) wide that forms a buffer between open seas and dense interior pack ice; here, waves, swells, and eddies have strong impacts that affect the ice, creating highly variable ice conditions.

Multiyear Ice—Ice that has survived at least one melt season; the thickness of multiyear ice floes can range from 2 to 20 m (6.6 to 66 ft) thick.

Predictability—The extent to which future states of a system may be predicted based on knowledge of current and past states of the system. Predictability is inherently limited because knowledge of the system's past and current states is imperfect and future variations of the external forcings are not exactly known.

Seasonal scale (21 days to 1 year)—The terms "prediction" and "outlook" are commonly used when referring to this timescale.

Statistical models—A model based on statistical relationships between different variables in past behavior of the system to be modeled.

Weather scale (1 hour to 10 days)—The term "forecasting" is commonly used when referring to this timescale.

sustained and coordinated conversation among the user, modeling, and observation communities. It was noted that this approach needs to go beyond ad hoc workshops and demands long-term, continuous, two-way interaction. This theme, which is discussed in more detail in Chapter 3, drove many of the key challenges and strategies laid out in the report.

The report addresses Arctic sea ice prediction over the seasonal to decadal timescales as a driver of the need for improved understanding of sea ice variability (Box 1.1). The committee's focus was on ice conditions during all seasons within the whole Arctic marine environment (i.e., Arctic Ocean and the subpolar seas, including the seasonal sea ice zone).

Although the Statement of Task does not explicitly mention stakeholders, it was the committee's view that a report on needs in sea ice prediction would be seriously deficient if stakeholders were not a prominent part of the underlying discussion. A similar sentiment was also raised in a recent NRC report: "IPY-related predictive modeling will continue to play a crucial role in helping commercial enterprises, individuals, and governments assess the regional and global risks associated with ongoing melting ice, sea level rise, permafrost degradation, and other effects of rising polar temperatures in a warming world" (NRC, 2012a).

Further, the committee and workshop participants observed that the motivational questions posed in the Statement of Task were not unique to this activity. Rather, they are questions that are often asked of researchers involved in observing and modeling the Arctic sea ice cover. This realization led the committee to consider additional, more overarching questions in the preparation of this report: (1) Given the significant investments and the progress that has been made in observing and modeling the Arctic sea ice cover, why are we not further advanced in the ability to predict its condition on seasonal to decadal timescales? (2) How can we apply the tools and insights we have developed in a systematic way to more effectively address the questions posed in the Statement of Task?

After presenting a series of key science questions, Chapter 2 identifies gaps and challenges related to understanding and predicting Arctic sea ice evolution. It begins with a set of overarching challenges that are foundational, including issues related to the Arctic environment and its stakeholders.

These overarching challenges are followed by challenges and gaps that are more specific to sea ice predictions, laid out as a function of timescale from seasonal to decadal. Chapter 3 presents possible strategies to significantly advance our understanding and predictions of Arctic sea ice over seasonal to decadal timescales. The organization of Chapter 3 is designed to generally follow the order of key challenges, though there is not a direct correspondence between the highlighted points made in Chapter 2. Examples of recent and ongoing activities are provided throughout Chapter 3 to demonstrate successful approaches that have been designed and implemented to address related issues. Key challenges and strategies are denoted in gray boxes throughout Chapters 2 and 3. Chapter 4 concludes with summary comments. Definitions of terms used throughout the report are provided in Box 1.2. This report does not make specific recommendations because of the reliance on the workshop in developing the ideas put forward in this report and the relatively short tenure for deliberations and analysis. The report does not include extensive background information. The interested reader is encouraged to utilize the numerous references and website links provided throughout the text.

2

Gaps in Our Understanding

In this chapter, the committee identifies a number of key challenges that currently limit our ability to understand and predict sea ice evolution. These key challenges were motivated by several key science questions that the committee considers crucial to answer in order to improve our understanding and ability to predict sea ice on seasonal to decadal scales (Box 2.1). The key science questions are intended to serve as a distillation of the more detailed discussions that follow.

OVERARCHING CHALLENGES

Many of the overarching challenges are not necessarily unique to the topic of sea ice prediction. However, their acknowledgment is important because they need to be considered in formulating strategies to advance predictive capabilities.

Key Challenge: Treating Sea Ice as Part of the Global System

A key to advancing our understanding and predictive capabilities is the treatment of the sea ice cover as an integral part of the complex Arctic system, which in turn is an integral element of the global system. Adding to this complexity is the broad range of human activities that not only influence the Arctic system, but are also influenced by it.

The Arctic sea ice cover is part of the larger Arctic system, which in turn is part of the larger global system, with interactions across all components. Arctic sea ice is a key element of these systems, influencing and influenced by change across a wide range of temporal and spatial scales (Figure 2.1). A myriad of feedback mechanisms link the atmosphere, sea ice, ocean, seafloor, and land, many of which are not yet fully understood (Maslowski et. al., 2012). For example, winds and ocean currents can alter the distribution of sea ice. These changes in the sea ice cover can then affect large-scale circulation patterns in the atmosphere (e.g., Deser et al., 2010; Francis and Vavrus, 2012) and ocean (e.g., Guemas and Salas-Melia, 2008), which in turn may impact weather, fisheries, and the global climate system. Other system components affecting and affected by sea ice cover include biological and chemical processes (e.g., Simpson et al., 2007; Kelly et al., 2010b; Deming and Fortier, 2011;). In addition, bathymetry plays a role in processes such as sea ice formation and evolution (Jakobsson et al., 2012; Nghiem et al., 2012).

BOX 2.1
KEY SCIENCE QUESTIONS

- What are the implications of the recent dramatic shifts in the Arctic from predominantly multiyear ice to first-year ice, and how will the associated complexities of this **regime shift** affect sea ice variability and predictability?
- In a rapidly changing Arctic regime, how will **forcings and couplings** between the various components of the ocean, atmosphere, cryosphere, and seafloor and land systems modify or influence the processes governing the characteristics of the sea ice cover?
- What are the impacts of **extreme events and feedback mechanisms** on Arctic sea ice evolution and our ability to predict it?
- How will the **changing Arctic sea ice characteristics and dynamics** affect stakeholders on a variety of timescales, including prediction requirements?

Adding to the complexity of the natural system is a range of human interactions. Humans not only influence the Arctic (e.g., impacts of oil spills, ship discharges, and land and sea greenhouse gas emissions), they are also influenced by it (e.g., decisions to pursue commercial shipping routes and offshore developments in response to a changing ice cover, long-term strategies for Arctic security, and development of future Arctic monitoring systems). The Arctic sea ice cover, therefore, cannot be viewed in isolation, as it is constantly responding to a host of regional and global forces, and these are in turn directly influencing the ice cover's seasonal and decadal presence or absence. Treating the Arctic as an integrated whole and a vital component of the global system is necessary to significantly advance our understanding of the sea ice cover as part of a complex system.

Key Challenge: Impacts of the Regime Shift of Arctic Sea Ice

Understanding how the recent regime shift in the Arctic sea ice cover, resulting in a significant reduction in the amount of multiyear ice compared with first-year ice, affects the processes governing the atmosphere-sea ice-ocean system and the models and observations used to study and predict Arctic sea ice dynamics.

The recent decline in the extent of Arctic summer sea ice has resulted in a dramatic shift in its composition (Figure 2.2). First-year ice is becoming more prevalent within the Arctic Basin, reducing the size of the multiyear ice pack (e.g., Maslanik et al., 2011). Furthermore, the multiyear ice that does remain is younger and thinner (Haas et al., 2008; Comiso, 2012). This rapid change to a new state is likely to have important implications for sea ice variability (Goosse et al., 2009; Hutchings and Rigor, 2012) and, ultimately, predictability (Holland et al., 2011). For instance, recent research suggests that first-year ice is not only more

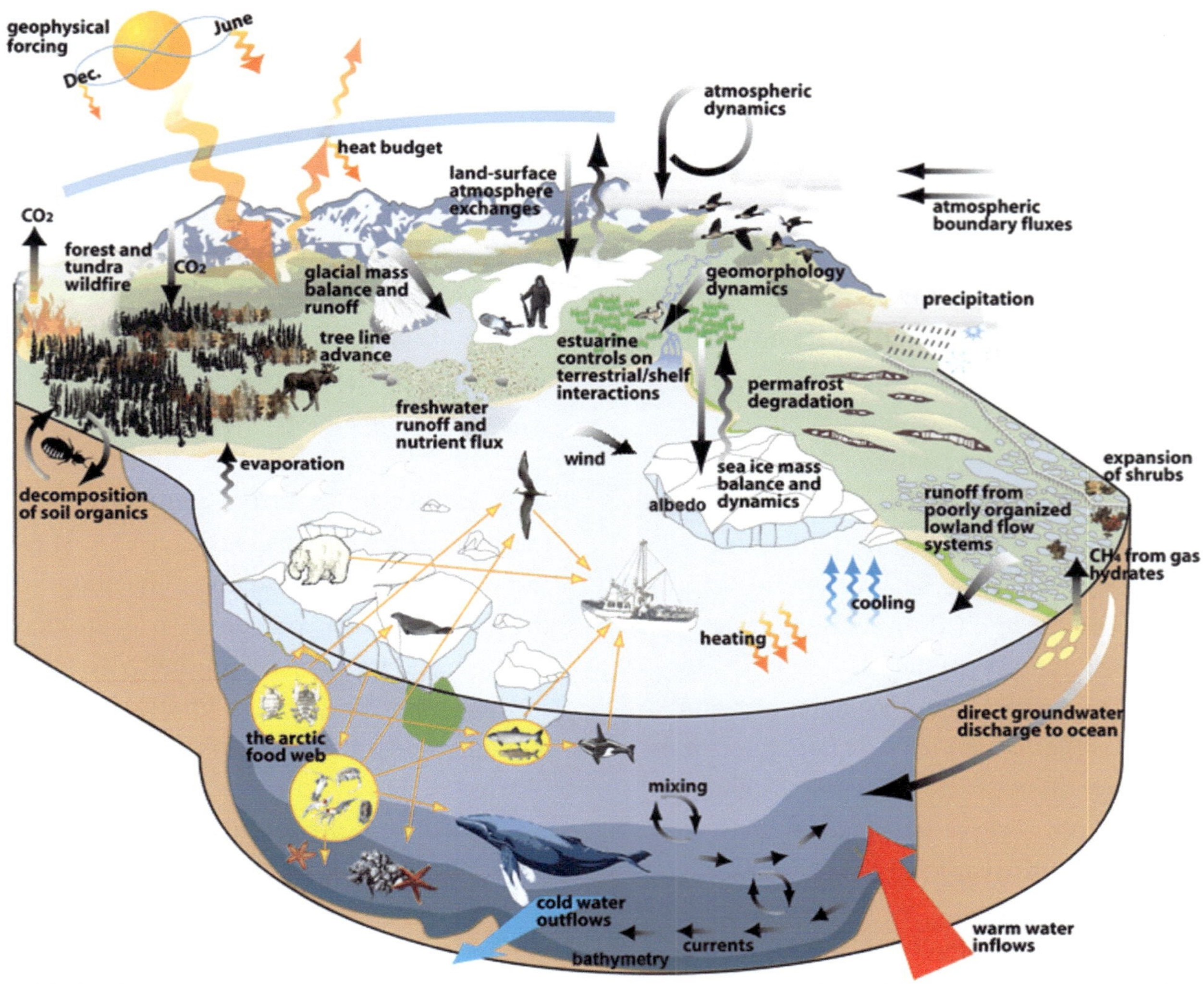

FIGURE 2.1 The Arctic system is made up of various components including a complex network of process interactions, interdependent feedbacks, and thresholds. There are many interconnections among system components and important changes in one component may influence numerous other parts of the system, including sea ice. SOURCE: Adapted from the Study of Environmental Arctic Change.

susceptible to summer melt (Perovich and Polashenski, 2012) but it is also likely to be more easily ridged, ruptured, and transported by winds (Rampal et al., 2009).The delay in the formation of shore-fast ice as well as its reduced stability and overall duration are also significant changes, with implications for stakeholders along the Arctic coast (Mahoney et al., 2007).

Some of the greatest changes to the Arctic Ocean are occurring in the Chukchi and Beaufort seas, where the increased summer ice retreat has created a substantially enlarged marginal ice zone (Holland and Stroeve, 2011). As a seasonal feature contained within the Arctic Basin, this region of transition between the ice-covered and open ocean is expected to have different governing processes than the marginal ice zones in the Greenland Sea or Southern Ocean. There are also indications that the extent of the North Atlantic marginal ice zone has increased over the past several decades (Strong, 2012).

It is important to understand how the transition from a multiyear to first-year ice regime affects (a) the realism of current numerical models and their parameterizations, (b) the power of statistical prediction methods, (c) the processes governing the atmosphere-sea ice-ocean system, and (d) instrument design and observational strategies. Currently, sea ice models' treatment of ice dynamics and thermodynamics employs parameterizations that were often developed using insights gained from observations taken in a primarily multiyear ice regime. For example, many sea ice models use ice dynamic and ice thickness distribution treatments based on the Arctic Ice Dynamics Joint Experiment that took place during the 1970s (e.g., Thorndike et al., 1975; Hibler, 1979). Model parameterizations based on the Arctic system's behavior in the past may not apply in the new state. Moreover, it is likely that if, as expected, the substantial ice retreat continues and the remaining ice transforms to a largely seasonal character, the oceanic and atmospheric circulation and thermodynamic structure will respond to the changes in the surface state, affecting large-scale patterns. The regime shift may also cause changes in physical and biochemical processes that are not or have not been adequately accounted for in current models. These complex interactions need to be simulated with sufficient accuracy to robustly project changes in the system.

A number of smaller-scale processes also need to be parameterized differently in first-year and multiyear ice because of inherent variations in their morphologies. For instance, recent work by Perovich and Polashenski (2012) conducted near Barrow, Alaska, found that once surface melt begins, seasonal ice albedos are consistently less than those for multiyear ice. This finding suggests that the shift from a multiyear to seasonal ice cover has significant implications for the heat and mass budget of the ice through a strong positive feedback on melting, and for primary productivity through an increase in the amount of sunlight penetrating into the upper ocean (Arrigo et al., 2012).

A shift from predominantly multiyear to mostly first-year ice will also affect the skill of statistical prediction methods. These methods use statistical relationships determined from past system behavior to predict the future state of the ice cover. An underlying assumption is that these relationships are stationary. However, given the transition to a thinner and increasing first-year ice pack, statistical relationships that have provided predictive skill for the past may no longer be valid.

The fundamental shift in the ice regime is also likely to have significant implications regarding the design of observation studies and networks, including the instrumentation. As is the case with sea ice models, most Arctic observational system designs and instruments were "developed" in the era of a predominantly multiyear sea

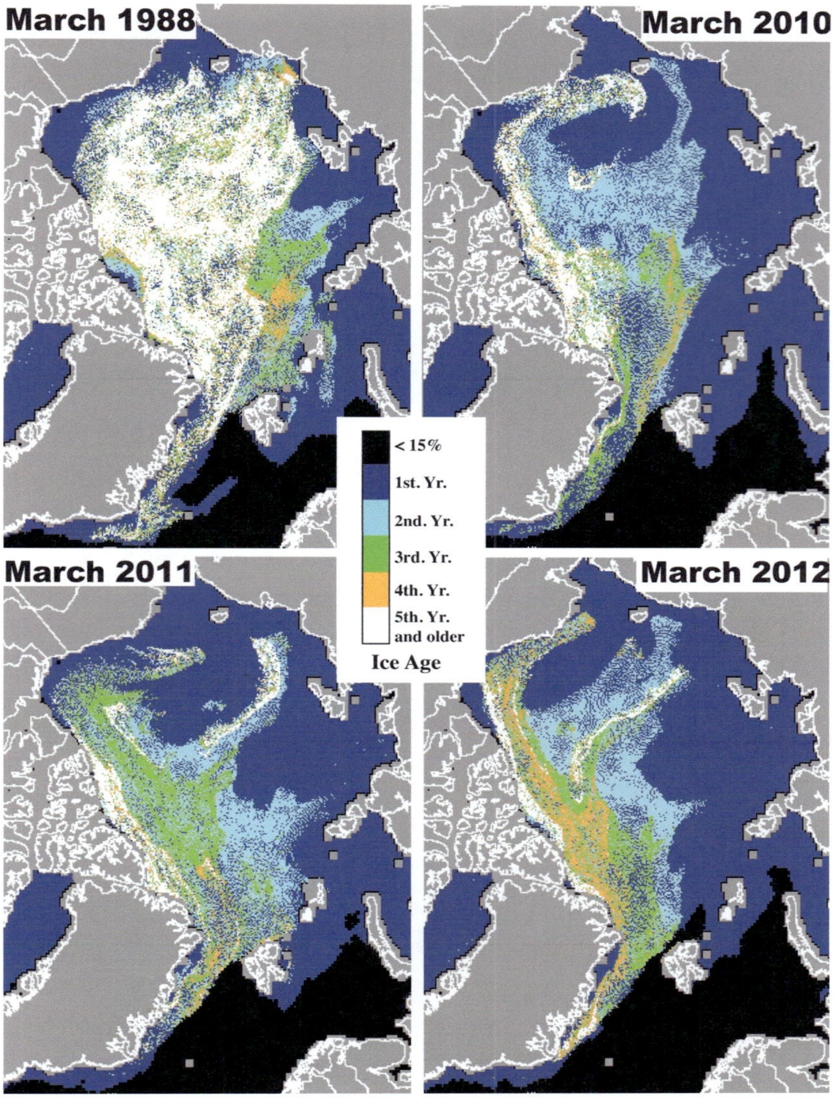

FIGURE 2.2 Sea ice is shifting from a composition of predominantly multiyear ice to include increasing amounts of first-year ice. Sea ice age distribution in March 1988 (upper left) compared with 2012 (lower right) illustrates the extensive loss in recent years of older ice types. Ice age is determined using satellite observations to track ice parcels. Older multiyear ice tends to be thicker and thus more resistant to forcing than younger first-year ice. SOURCE: James Maslanik, Charles Fowler, and Mark Tschudi.

ice cover and thus may need retooling to address thinner, more mobile ice conditions.

For example, many buoys deployed in coordination with the International Arctic Buoy Programme were largely designed for placement on multiyear sea ice floes to maximize their longevity. There are recent examples of work dedicated to transitioning the design of ice-based instruments for deployment in the seasonal ice zone, where they tend to be more vulnerable (e.g., Polashenski et al., 2011). New sustained measurement strategies can also take advantage of innovative technologies such as unmanned aerial systems and autonomous underwater vehicles to answer specific questions without the risks of deployment on seasonal sea ice or in open water.

Key Challenge: Identifying Diverse and Emerging Stakeholder Requirements

As the Arctic is being transformed by globalization and climate change, new stakeholders with additional and more sophisticated requirements are emerging. Clearly defining these diverse needs as they relate to seasonal to decadal sea ice prediction is crucial to inform the future directions of modeling, observations, and overall research.

The term stakeholder covers a wide range of communities with interests in the development and application of Arctic sea ice prediction across seasonal to decadal timescales. The term includes indigenous residents on the Arctic coast, scientific researchers, commercial users (e.g., fishing, shipping, natural resource development, and marine tourism.), and naval and coast guard planners. Not surprisingly, this diverse group of stakeholders has a broad and evolving set of needs and, hence, sea ice prediction requirements. In addition to ice type (e.g., multiyear/perennial vs. first-year/seasonal ice), extent, thickness, and motion, there is growing interest in the ability to predict snow cover depth, melt onset, freeze-up/ice season length, and the size and shape of ice floes (e.g., Ray et al., 2010). For example, earlier snowmelt onset date and thinner snow cover during winter are important parameters for an early snowmelt, which may prematurely destroy subnivean lairs subjecting ringed seal pups to adverse weather and increased predation (Kelly et al., 2009; Hezel et al., 2012; see Figure 2.3).

Understanding stakeholders' needs for seasonal to decadal sea ice prediction is crucial to guiding the future directions of modeling, observations, and overall research. Effectively quantifying and communicating the inherent limitations in sea ice predictability is also needed to establish reasonable stakeholder expectations. Significant issues that will need to be overcome include: (1) how to aggregate stakeholder input; (2) how to prioritize among stakeholder interests and needs; (3) how to communicate the limitations of modeling and forecasting to users and policy makers; (4) how to maximize the utility of sea ice predictions containing uncertainties; and (5) how to assess opportunities for overlapping interests and efficiency of efforts in meeting stakeholder needs.

A realistic expectation for the future state of Arctic sea ice is needed to manage risks, deploy limited resources, and plan long-term infrastructure. Plausible future sea ice scenarios projected by fully coupled, large-scale global climate models (GCMS) can be useful for strategic planning. For instance, they are used by federal and private entities and commercial shipping companies looking to understand future navigation seasons on potential Arctic shipping routes. GCMs are also used by federal agencies to study the plausible futures of endangered

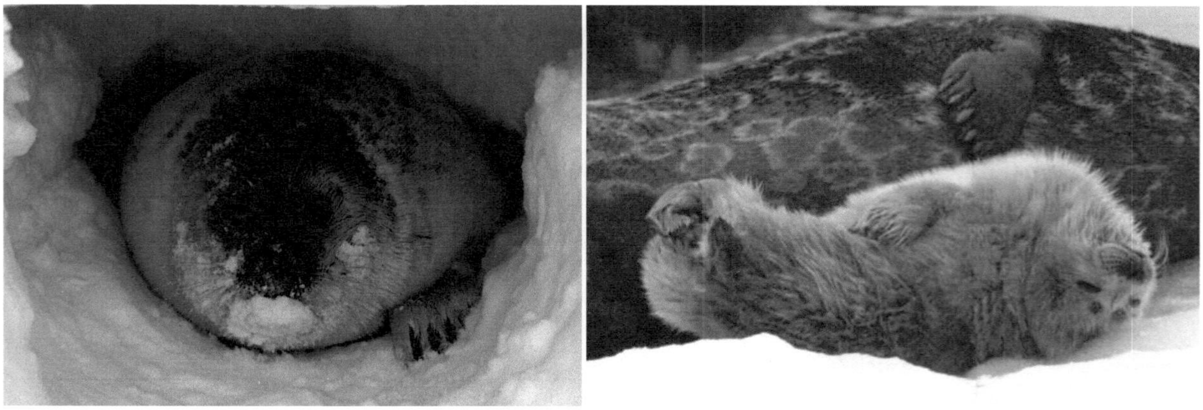

FIGURE 2.3 Where sufficient snow accumulates on the sea ice cover, ringed seals may excavate snow caves (lairs) above the breathing holes to protect the pup from adverse weather and predation. Early snowmelt onset causes premature lair abandonment exposing pups to hyperthermia and predation. Images courtesy of Brendan Kelly (left) and Juha Taskinen (right).

species and the possible merits of regulatory strategies (e.g., NOAA, 2010). However, better characterizing the uncertainties through continued intercomparisons of model outputs as well as validation with sea ice observations is necessary to enhance confidence in using this information for decision making.

Although projections from GCM simulations fulfill some requirements for planning, regional forecasts and seasonal predictions at shorter timescales and finer spatial resolutions are more useful to offshore drilling operations, marine transportation systems, and subsistence hunters. Many of these requirements call for timely predictions that resolve the highly varying ice characteristics that occur near the coastline, demanding accuracy of the prediction within a few hundred kilometers (or tens of kilometers) of shore and within the marginal ice zone. Modelers are developing methods to address these needs—by either nested grids or regional high-resolution models—but they face challenges related to the extreme computational demands, lack of suitable initialization conditions, inadequate projections of atmospheric and oceanic forcing, and insufficient software and resources to manipulate and store the enormous datasets produced. Parameterizations that have been developed for relatively coarse resolutions, such as the viscous-plastic sea ice rheology (Hibler, 1979), may not be applicable at very fine spatial scales. This may necessitate additional model developments such as the incorporation of more detailed representations of sea ice dynamic components (e.g. Hopkins and Thorndike, 2006; Sulsky et al., 2007) into large-scale climate models.

CHALLENGES IN ADVANCING PREDICTIVE CAPABILITY

Key Challenge: Competing Approaches to Seasonal Sea Ice Prediction

Although limitations in the various approaches used to generate seasonal forecasts are generally acknowledged, there is a lack of quantitative information about their relative strengths and weaknesses.

Three approaches are used to provide seasonal ice forecasts, including (1) statistical algorithms (e.g., Drobot et al., 2006); (2) coupled ice-ocean models driven by prescribed atmospheric forcing (e.g., Zhang et al., 2008b); and (3) more recently, coupled atmosphere-ocean-ice models. All three approaches have contributed to the SEARCH (Study of Environmental Arctic Change[1]) Sea Ice Outlook.[2] Each of the methods has acknowledged limitations, but the relative accuracies, strengths, and weaknesses are poorly known. For example, statistical relationships based on past observations may not be valid in a future Arctic dominated by thinner ice (Holland and Stroeve, 2011).

In coupled ice-ocean prediction systems, atmospheric forcing is usually based on an ensemble of past years' winds and temperature fields derived from atmospheric reanalyses. However, the statistics of past forcing fields may not be applicable to the new Arctic state, because the preponderance of thin ice, large-scale changes in surface characteristics, and resulting shifts in atmospheric patterns all affect surface-atmosphere exchanges. Additionally, important feedbacks between the atmosphere and ice that influence the sea ice evolution are not accounted for by using historical atmospheric conditions. In coupled atmosphere-ocean-ice model predictions, atmospheric forcing is free to evolve consistent with the underlying distributions of sea ice and ocean temperatures, but the simulations of sea ice evolution suffer from systematic biases. This causes the models to drift to their preferred (and biased) climate state once constraints by observations are removed. Methods to bias correct (or "de-drift") forecasts with coupled models have been explored (e.g., Yeager et al., 2012, for ocean conditions). Alternatively, "anomaly-initialization" in which observed anomalies are added to the model"s preferred climate state has been used in a number of decadal prediction studies (e.g., Pohlmann et al., 2009).

The relative benefits of these various approaches have not been assessed for sea ice prediction. Determinations of their prediction accuracy are challenged further by observation-based uncertainties in the sea ice variables that are to be predicted. This is especially the case when the predictions extend to new variables (e.g., sea ice thickness, morphology, and stability) that are of particular and growing interest to stakeholders.

Key Challenge: Observational Requirements for Seasonal Sea Ice Prediction

[1] http://www.arcus.org/search/index.php.

[2] http://www.arcus.org/search/seaiceoutlook/index.php.

Seasonal sea ice prediction capability depends on adequate knowledge of initial ice-ocean conditions, even though the specific requirements associated with "adequate knowledge" have yet to be established. This challenge is compounded by the need for a fast turnaround in acquiring and accessing observational data.

There has been a dearth of experiments performed to systematically evaluate the sensitivities of sea ice predictions to the type, quality, density, and frequency of observations of the Arctic atmosphere, ocean, and sea ice. These sensitivities arise from the specification of the initial state of the sea ice and its drivers in numerical models.

During the forecast period, atmospheric variations, which are largely unpredictable beyond a week or two, are expected to have a significant influence on sea ice and will limit its seasonal predictability. Instead, sea ice predictability on these timescales resides in the initial ice and ocean state. Factors believed to be among the most important for predicting sea ice behavior on the seasonal scale include accurate knowledge of (1) sea ice conditions at the start of the season (particularly the ice thickness distribution and the partitioning between seasonal and multiyear ice) and (2) upper ocean conditions, such as ocean mixed-layer heat content.. However, the relationship between prediction skill and the uncertainty in each of these factors is poorly known. The importance of initial values of other variables—such as snow on sea ice, ocean temperature and salinity profiles below the ice, and ocean current distributions—is poorly understood, but may be considerable. The accuracy requirements for bathymetry, which constrains ocean currents and controls the distribution of warm and cold water masses, are largely unknown (Jakobsson et al., 2012).

There is little information on what observational quality, spatial density, location, and accuracy are required for different variables to realize useful predictive power. Compounding the problem is the fact that observations of the atmosphere, sea ice, ocean, and seafloor are made from vastly different sensors, including in situ, airborne, and satellite instruments, along with a variety of methods. Each approach brings with it a specific set of characteristics, which often go undocumented.

There has been little if any effort to establish and broadly apply measurement protocols and standard definitions. For instance, even for one of the most fundamental parameters such as air temperature, there is no existing national or international protocol for predeployment calibration, maintenance of measurement stability during deployment, post-deployment calibration, and cross calibration among different sensors and different algorithms (e.g., different algorithms to measure ice surface temperature from different satellite sensors).

Adequately acquiring and making observations available within several days of the beginning date of the model simulation are necessary if the observations are to be useful for initializing operational seasonal model predictions. Ice extent is available within a day of satellite data acquisitions from multiple satellite sensors (passive microwave, active microwave, and multispectral sensors). Recent efforts undertaken by the National Aeronautics and Space Administration (NASA) IceBridge project (Kurtz et al., 2012) and the Sea Ice Outlook,[3] for example, have demonstrated the ability to deliver observations of sea ice thickness and snow depth within a period of weeks from airborne platforms. This achievement was the result of a coordinated effort backed by dedicated resources (i.e., measurement platform, instruments, personnel, and funding resources).

However, near-real-time access remains problematic for many other types of data, such as ocean temperatures, salinities, and ice draft collected from seafloor moorings. Moreover, ocean currents, the distribution of warm and cold waters, and ocean mixing are controlled, guided, or restricted by seafloor characteristics. About 89 percent of the Arctic Ocean still needs to be surveyed to determine where interpolation or extrapolation may result in unrealistic bathymetric features (e.g., false seamounts) or where important structures (e.g., canyons) have been missed (Jakobsson et al., 2012). In general, prioritizing data acquisition with sufficient temporal and spatial coverage and timely accessibility to observations is critical if they are to play an effective role in seasonal, operational sea ice prediction.

Coupled climate models calculate their own atmospheric and oceanic variables from the basic laws of physics, providing the thermodynamic and dynamic drivers of sea ice evolution. Considering the number of variables and interactions that affect the ice cover, it is a tremendous achievement that the models simulate the observed ice behavior as well as they do. Any errors in the variables, processes, or feedbacks involved in these calculations, however, will be propagated and integrated by the ice over time and result in an unrealistic representation. A key challenge is simulating realistic atmospheric and oceanic conditions, which in turn depend on assumptions about future trends in carbon dioxide emissions, aerosol loading, changing surface characteristics, etc. Related to this challenge is the difficulty in determining which processes in a particular model are responsible for unrealistic aspects of sea ice simulations, especially systematic biases such as those responsible for the large model spread in Figure 1.3.

Key Challenge: Projecting Realistic Forcings and Feedbacks for Decadal Sea Ice Predictions

A key challenge in coupled climate models is the capability to realistically simulate atmospheric and oceanic conditions of relevance to sea ice variability, including the identification of model processes that contribute to unrealistic forcings and feedbacks.

Phenomena such as El Niño-Southern Oscillation (ENSO), the Arctic Oscillation (AO), the Pacific Decadal Oscillation, and the Arctic Dipole are known to affect the ice cover through their influence on ice transport, storm tracks, and heat transport (e.g., Rigor et al., 2002; Wang et al., 2009; Ogi et al., 2010). What drives these

[3] http://www.arcus.org/search/seaiceoutlook/ice-thickness-data.

atmospheric patterns into their positive and negative phases in any given year or sequence of years, however, is not well understood. Although models generally simulate these major modes of variability, studies reveal potentially important discrepancies in the statistics of their variations (e.g., Stenchikov et al., 2006). Recent studies have suggested that phasing of the AO may be related to stratospheric influences in some situations (Black, 2002; Cohen and Jones, 2011) and to surface forcing in others (Overland and Wang, 2010). Thus models need to include a realistic representation of the stratosphere and its interaction with the tropospheric circulation.

The present capability to predict, in detail, the large-scale modes of atmospheric variability is limited, and in some cases there may be little deterministic predictability beyond a few weeks. Experiments addressing such changes in large-scale atmospheric modes are challenged by natural variability, because many ensemble members can be required to detect significant changes in the circulation (Bhatt et al., 2008; Deser et al., 2012). Despite these challenges, models need to realistically simulate the statistics of these atmospheric variations, including their spatial patterns, frequency of occurrence, and response to varying forcings, if they are to accurately simulate sea ice variability on decadal scales. Whether fluctuations in spatial extent, intensity, and frequency of these large-scale oscillations will change as greenhouse gases continue to accumulate is a key open question. Evidence does suggest that high-latitude surface changes—which include changes in sea ice cover that may affect wind patterns (e.g., Liu et al., 2012), water vapor content, and cloud amount (e.g., Winton, 2006; Kay et al., 2012)—can then feed back onto the ice cover (Overland et al., 2012;Wu et al., 2012) and steer ocean currents. Capturing these feedbacks in coupled models is critical if decadal predictions are to be successful.

As evidenced in summer 2007, extreme events (e.g., anomalous winds as one of several key factors) may combine with preconditioning and ice-albedo feedback to result in abrupt change (e.g., a large decrease of sea ice in a short time) (Haas et al., 2008; Perovich et al., 2008; Zhang et al., 2008a; Lindsay et al., 2009; Ogi and Wallace, 2012) that can have decadal impacts. For example, drastic loss of perennial sea ice owing to persistent wind patterns in 2005 and 2007 (AMAP, 2011) may influence the long-term sea ice trends. Models do simulate extreme events of this type (e.g., Holland et al., 2006) but the realism of how simulated extreme events modify key parameters needs to be further assessed.

Another characteristic that highlights the interconnectedness of the Arctic system is the influence of the Atlantic and Pacific oceans on the Arctic Ocean. Relatively warm water masses from the Atlantic and Pacific enter the Arctic Ocean, and because they are saltier than the surface waters, reside below the mixed layer. The Arctic Ocean's present stratification, resulting primarily from the vertical salinity profile, largely limits heat transfer to the ice cover from the deeper layers. These deeper layers contain vast quantities of heat that could melt all of the sea ice relatively quickly (e.g., Alexeev et al., 2011). As the Arctic transitions to a state dominated by first-year ice and as atmospheric circulation patterns potentially change in response to sea ice loss and other factors, the stratification of the upper-ocean layers could also change, modifying the amount of Atlantic and Pacific layer heat that affects the ice cover. Moreover, additional solar energy absorbed in areas where ice has retreated will also affect mixed-layer characteristics (Perovich et al., 2007). Owing in part to insufficient vertical resolution in modeled oceans, processes related to the mixed layer and stratification are not well captured by models. Additionally, lateral oceanic transports affecting, for example, the net heat flux from the North Atlantic into the Arctic Basin, are also often poorly simulated in current climate models.

3
Strategies for the Future

OVERARCHING STRATEGY

Key Strategy: A Deliberately Integrative Approach for Sustained and Coordinated Collaboration among User, Modeling and Observation Communities

A deliberately integrative approach is needed to facilitate coordinated and sustained discussions and collaborations among the user, modeling, and observation communities to inform effective research activities and to set realistic expectations for predictions. Such an approach would need to take full advantage of existing infrastructure and draw from comparable efforts in other fields.

The need for sustained and facilitated communications among the many stakeholder communities is both a challenge and a significant opportunity. End-user needs are important drivers for science and research efforts and for more accurate, timely and useful sea ice forecasts. For example, regularly scheduled, iterative, and sustained discussions among end users, modelers, field scientists, and the remote sensing community could help the science communities understand user tolerances for the accuracy of particular ice characteristics. They could also help to determine the most practical approaches to obtaining the highest priority products.

Some user requirements may be beyond existing capabilities because of inherent limitations in predictability or owing to technical or practical limitations. Thus, it is incumbent upon the research element of the stakeholder community, including modelers and observers, to understand the basis of these requests, to explore alternatives, and to properly manage expectations. Increased collaboration can help overcome the challenge of communicating to the public about what can reasonably be expected from seasonal to decadal sea ice prediction.

Within the framework of organized stakeholder communication, true collaborations between the modeling and observation communities are essential. Given the reality of limited resources, these collaborative efforts are more likely to identify resource needs and shared resources. The need for integrative, sustained conversations calls for routine and frequent engagement. Although workshops and topical meetings have their place within the process, this strategy calls for a longer-term, continuous interaction. The

BOX 3.1
Examples of Sustained Conversations: ArcticNet, Alaska Ocean Observing System (AOOS), and Alaska Center for Climate Assessment and Policy (ACCAP)

An example of a large-scale program that facilitates sustained communication across the research and stakeholder communities is ArcticNet,[a] a network of Canadian institutions that aims to bring together scientists and representatives from academia, government, industry, international agencies, and northern communities. The main objective of ArcticNet is to utilize information from these sources to help Canada prepare for and adapt to climate change and increased activity in the Arctic. Through research collaborations and partnerships with corporations operating in the Arctic, reliable scientific data can be made available to a wide variety of stakeholders. Research results are shared with the community in the form of scientific publications and a data catalogue and through various media outlets. This integrated and cross-disciplinary approach to Arctic research provides unique opportunities for education, collaboration, and entrainment of the next generation of polar researchers.

The Alaska Ocean Observing System (AOOS[b]) is another example of a network intended to address both national and regional needs for researchers and stakeholders on coastal and ocean issues. In coordinating federal, state, local, and private needs, AOOS identifies gaps in data, helps fill those gaps when appropriate, and increases the usefulness of existing data. AOOS demonstrates the integration and collaboration necessary to enable a variety of users to obtain information and to make decisions about the marine environment in the Alaska region.

mechanisms to support and facilitate these sustained conversations would need to be deliberately identified and implemented, as opposed to relying on self-organization. This would require close and effective engagement among public, private, and academic institutions. Participants in these conversations could serve a role akin to that of a diplomat, seeking and communicating ideas and suggestions that reflect a broad viewpoint. There are excellent examples of efforts underway, on both national and local levels, to develop and facilitate interactions among relevant research communities and users (Box 3.1).

Looking more specifically at the issue of sea ice prediction, one example of an initial framework for capturing stakeholder needs in terms of variables utilized in sea ice science is provided in Table 3.1. Other frameworks can be used to organize user-science connections with the goal of advancing the utility of sea ice predictions. For this process to be successful, it is important for communities to learn each others' language and to be aware of usage differences. For instance, the terms sea ice "memory" and "persistence" have specific meanings within the sea ice research community that may not be appreciated by

Information is made available through various data management and information products (including a website and data portal), workshops and reports, and newsletters. AOOS also makes educational resources available for teachers and interested community members. The AOOS network includes mariners, fishermen and subsistence users, search and rescue operations, scientists, coastal security operations, resource managers, and educators.

A regional-scale example of a forum through which stakeholders work directly with scientists is the Alaska Center for Climate Assessment and Policy (ACCAP[c]). The goal is to enable Alaskans to be able to respond to climate changes by targeting products (e.g., National Oceanic and Atmospheric Administration [NOAA] forecasts) to address specific user needs. Using tools such as webinars, video conferences, web-based guides and maps, and social media, ACCAP reaches out to encourage dialogue between scientists and end users. These resources help convey valuable information on climate change science as well as information on uncertainty and risk management. Strategies and plans to adapt to climate change are developed in coordination with stakeholders, agencies, industries, and citizens to ensure that partnerships are built and information needs are met. The ACCAP program is part of the NOAA Regional Integrated Sciences and Assessments (RISA[d]) program, which supports research on complex, interdisciplinary issues that are addressed at the regional level.

[a] www.arcticnet.ulaval.ca/index.php
[b] www.aoos.org/
[c] http://ine.uaf.edu/accap/
[d] www.climate.noaa.gov/cpo_pa/risa/

others. Even commonly used terms such as "multiyear" sea ice may have different meanings that need to be clearly communicated among the relevant stakeholders.

These ongoing efforts and suggested framework offer important building blocks to advance the strategy envisioned here. However, in the committee's view this activity will likely not be effectively facilitated without a dedicated and deliberate effort backed by sufficient resources, including designated funding. For example, an NRC report noted that maintaining some networks developed and cultivated during IPY has been difficult and many of its valued components—such as the international IPY website, its publication database, and educational and outreach efforts—have struggled to find alternative resources (NRC, 2012a). The characteristics of these sustained conversations suggest leadership from a high-level, inter-governmental office, agency, or consortium.

TABLE 3.1 Example of an initial framework that could be useful in determining key stakeholder needs and prioritizing variables to be measured by the research community to better meet those needs.

Stakeholder Activity	Ice edge location	Ice concentration	%FY/ MY	Mean FY thickness	Mean MY thickness	Ice growth rate	Ice melt rate	Floe size	% Ridged ice	% Leads mean width	Snow depth	Melt/ freeze onset	Fast ice break-up date	Ice velocity	Ice character	Local	Reg-ional	Pan-Arctic
Infraructure siting																		
Commercial fishing																		
Marine Tourism																		
Shipping																		
National Security																		
Indigenous use																		
Science campaigns																		
Science applications																		
Marine Mammals																		
Offshore resource extraction																		
Coastal operations																		

NOTE: The top row lists sea ice variables and measurements. Appropriate cells could be filled with check marks or indications of priority. If such a table is completed by various stakeholders, a particular variable (e.g., floe size on seasonal timescales) for which a large number of stakeholders indicate a need may be given higher priority than a variable needed by fewer stakeholders. This table also shows the breadth of stakeholders and their needs. It can be used for both seasonal and decadal timescales.

SOURCE: Generated from discussions at the workshop.

STRATEGIES TO IMPROVE SEA ICE PREDICTIVE CAPABILITIES: SEASONAL TO DECADAL TIMESCALES

While recognizing that there are limitations in current modeling and observational techniques, the committee offers possible strategies to significantly enhance our understanding and predictions of Arctic sea ice cover over seasonal to decadal timescales. Implementation of these proposed strategies will require iterative interaction between model development and observational input, balanced by a sustained dialogue with end users.

Key Strategy: Evaluation of Existing Seasonal Prediction Methods

A coordinated and detailed comparison of the different approaches used to generate seasonal sea ice forecasts could establish baseline expectations for predictive skill and identify priority needs, setting the stage for advances in predictive capability.

As previously discussed, several methods are used to obtain seasonal forecasts of Arctic sea ice, including (1) statistical methods, (2) ice-ocean models driven by prescribed atmospheric forcing, and (3) fully coupled atmosphere-ocean-ice models. A coordinated comparison of these prediction methods will serve to inform both the science and the needs of stakeholders. It is important that the evaluations be based on regional metrics and not be limited to ice coverage. For instance, other candidate metrics are dates of ice retreat and closure of various sea routes or specific coastal locations. Various comparison approaches need to be considered, such as hindcast and so-called "perfect-model" studies.

In a hindcast model study, a retrospective assessment of past years, both initial conditions and validation data are needed. An evaluation of this kind was recently performed on forecasts of the El Niño-Southern Oscillation phenomenon (Barnston et al., 2012; Box 3.2). In the case of sea ice prediction experiments, initialization will be limited by the accuracy of key predictor variables (e.g., ice thickness), but such limitations will be common to all three approaches. Perfect-model studies can also provide useful insights to predictability. These studies treat simulation output from a coupled numerical model reference experiment as the "truth" (i.e., equivalent to observations). The different prediction methods discussed above can then be applied to forecast the conditions from this reference simulation. This can be performed for both past and future model-simulated conditions, allowing for information on how predictability characteristics may change with changes in the climate state. Such studies have the advantage of a complete knowledge of initial and time-varying conditions and provide the ability to address the implications for possible nonstationarity in statistical relationships for future climate states. It is important to remember that the results from these studies need to be considered within the context of the imperfect coupled model being used.

BOX 3.2

COMPARATIVE EVALUATION OF SEASONAL FORECASTING METHODOLOGIES FOR ENSO

Given that the El Niño-Southern Oscillation (ENSO) affects climate and weather events (e.g., drought, flooding, and tropical storms) on a global scale, understanding and improving forecasts are critical to both the scientific community and the public (McPhaden et al., 2006). Real-time ENSO prediction capability during the 1990s was assessed to be somewhat useful (Barnston et al., 1994), with dynamical and statistical models showing comparable skills.

Although there has been progress in the prediction of ENSO during the last decades (Randall et al., 2007), numerous issues regarding its dynamics, impacts, and predictability remain uncertain. In the last two to three decades, the ability to predict warm and cold episodes of ENSO at short and intermediate lead times has gradually improved due to:

- Improved observing and analysis/assimilation systems,
- Improved physical parameterizations,
- Higher spatial resolution, and
- Better understanding of the tropical oceanic and atmospheric processes underlying the ENSO phenomenon (e.g., Guilyardi et al. 2009).

Findings from a more recent study on ENSO predictions suggest that additional advances in capabilities are likely with the expected implementation of better physics, numeric and assimilation schemes, finer resolution, and larger ensemble sizes (Barnston et al., 2012). The study evaluated real-time model predictions of ENSO conditions during the 2002-2011 period and compared them with skill levels documented in studies from the 1990s. The skills of 2002-2011 models is slightly better than that of earlier decades, with the recent decade's dynamical ENSO prediction models outperforming their statistical counterparts to a slight but statistically significant extent. The greater power of dynamical models is largely attributable to the subset of dynamical models with the most advanced, high-resolution, fully coupled ocean-atmosphere prediction systems using sophisticated data assimilation systems and large ensembles (Barnston et al., 2012).

Key Strategy: Process-Based Studies Targeted at the Increasingly Prevalent First-Year Ice Cover

Questions surrounding the impact of the trend toward an increasingly seasonal Arctic sea ice cover could be addressed with the development of a highly coordinated and integrated process-based study, analogous to the Surface Heat Budget of the Arctic Ocean (SHEBA) project, focused on understanding oceanic, atmospheric, and terrestrial contributions to seasonal sea ice predictions.

The fundamental properties of the ice cover are changing as the Arctic transitions toward a seasonally ice-free state, resulting in a significant reduction in the amount of multiyear ice compared with first-year ice. In the face of this significant transition, there is the need to identify and understand whether and how key parameters (i.e., first-order effects) influence predictability. A likely outcome is the need for improved model formulations of the dynamic and thermodynamic processes governing the behavior of a sea ice cover composed of largely first-year ice.

Previous work done on the fundamental properties of first-year sea ice (e.g., Weeks and Ackley, 1982; Timco and Weeks, 2010) can inform the design of process studies that will advance our understanding of first-year ice in a predictive context. The challenge lies in developing a thorough understanding of the fundamental properties of first-year sea ice act together on a large, pan-Arctic scale to affect air-ice and ocean-ice heat transfer (ice thermodynamics) and ice pack mobility (ice dynamics, e.g., Melling et al., 2005; Amundrud et al., 2006; Barber et al., 2012). Moreover, as the sea ice cover becomes dominated by the weaker and less stable first-year sea ice, it may be more susceptible to extreme events that could have impacts lasting from seasonal to decadal timescales.

Observations of these processes and their interactions are needed to determine which aspects of existing predictive models require further development and to help determine requirements for sustained observations necessary to verify model realism as the ice cover continues to evolve. Conversely, model experiments can identify which process parameterizations are the greatest sources of uncertainty (or error) in climate model simulations. Systematic sensitivity experiments, performed with an ensemble of different models, would initiate an end-to-end process study in which the needs of models guide field experiments, which in turn feed back (via improvements in process formulations or parameter estimates) to models used for sea ice prediction.

An end-to-end process study in the seasonal ice zone, guided by past work, historical data and the output from sensitivity studies using current models, would enhance process understanding and simulation capability. Although process understanding is especially important for predicting the evolution of initial conditions over seasonal and interannual timescales, it also has potential for multiyear timescales, based on the apparent multiyear timescales of the ocean inflow anomalies, at least in the Atlantic sector (Polyakov et al., 2010). The nature of a process-based study of

BOX 3.3
Examples of SHEBA-Like Initiatives

There are a few initiatives under way that could be built upon and possibly integrated to conduct a SHEBA-like project in seasonal ice:

- **Office of Naval Research 5-year Emerging Dynamics of the Marginal Ice Zone Department Research Initiative[a] (ONR MIZ-DRI).** The 5-year ONR MIZ-DRI was initiated in 2011 and includes plans for a field project in 2014. The observational dataset that is generated from an integrated suite of platforms can be used to evaluate the skill of numerical forecast models.

- **Ocean Observations to Improve Sea Ice Forecasting.** This project was initiated during the Arctic Observing Coordination Workshop, held March 20-22, 2012, in Anchorage, Alaska[b]. In its very early planning stages, the project is designed to provide the necessary ocean observations to improve sea ice forecasting on daily, seasonal, interannual and decadal timescales.

- **Multidisciplinary Drifting Observatory for the Study of Arctic Climate (MoSAIC).[c]:.** The concept of MoSAIC is to establish an international, multiyear, manned, drifting observatory in the central Arctic sea ice pack to obtain observations of atmosphere, sea ice, and ocean processes that will compose a testbed for process, regional, and global model evaluation and development.

atmosphere-sea ice-ocean coupling in the seasonal ice zone will almost certainly require a set of observations over a full annual cycle. The process study is also inherently interdisciplinary because of the broad science scope encompassing surface-to-satellite observations and models extending across the atmosphere, ocean, seafloor, and land.

The year-long Surface Heat Budget of the Arctic Ocean (SHEBA) project (Perovich et al., 1999), conducted from 1997 to 1998 in multiyear ice, serves as an excellent model for a study of this kind. SHEBA was designed to facilitate interactions between observation and modeling research communities, through the planning, implementation, and analysis phases of the project. Now that it has been over a decade since the project formally concluded, it is apparent that one of the major successes of SHEBA was the interdisciplinary teamwork that brought together a diverse group of researchers, each bringing their own particular expertise, to work on the common goals of the program (Perovich et al., 2003).

SHEBA continues to motivate cross-cutting collaborations that advance our

- **Year of Polar Prediction[d] (YOPP):** This is a joint effort from the World Weather Research Progamme and the World Climate Research Programme. The YOPP is tentatively scheduled for 2017-2018 and will include an intensive observation and modeling period to provide data for model development, data denial experiments, and predictability and diagnostic studies.

Other examples of successful studies of seasonal ice in the past 15 years include:

- **International North Water Polynya Study (NOW):** This program was initiated to study the North Water polynya and the mechanisms associated with its formation and biological production.
- **Canadian Arctic Shelf Exchange Study (CASES):** This was an international effort under Canadian leadership to understand the biogeochemical and ecological consequences of sea ice variability and change on the Mackenzie Shelf.
- **Circumpolar Flaw Lead (CFL) Study:** The objective of this project was to examine how physical changes affect biological processes within the flaw lead.

[a] www.onr.navy.mil/en/Science-Technology/Departments/Code-32/All-Programs/Atmosphere-Research-322/Arctic-Global-Prediction/Marginal-Ice-Zone-DRI.aspx
[b] www.arcus.org/search/meetings/2012/coordination-workshop
[c] www.esrl.noaa.gov/psd/events/2012/mosaic/
[d] www.wmo.int/wwrp

understanding of the processes governing sea ice thermodynamics. This significant and enduring outcome suggests that the investment in a large, focused effort, if effectively coordinated and implemented, can have a greater impact than a more diffuse approach. If followed, a key addition to the approach used during SHEBA would be the increased involvement of stakeholders outside of the sea ice research community, including a greater emphasis on the Arctic marine ecosystem, atmospheric chemistry, the coastal terrestrial system, and, more generally, end users (e.g., indigenous populations, natural resource industries, fishing communities, commercial shippers, and marine tourism operators). Some examples of SHEBA-like initiatives can be found in Box 3.3.

A comprehensive process study in the seasonal ice zone also offers the opportunity to identify, develop, and test instruments and observational platforms that can effectively and efficiently support both seasonal and decadal prediction capabilities. Measurements of sea ice motion, ice thickness, and snow cover/depth are made from a variety of sensors—including in-situ, airborne, and satellite instruments—each having different capabilities. Existing surface networks (e.g., buoys and weather stations) will need to be sustained and extended with aircraft campaigns and new instruments. Continuous satellite observations (e.g., multispectral sensor, synthetic aperture radar, passive microwave radiometer, active microwave scatterometer, and altimeter) over decadal timescales are critical to assess the role of sea ice change in the global climate system.

To enhance predictive capability, the process study needs to have a focus on improving predictions on seasonal to decadal scales. It also needs to involve stakeholders and modeling and observational communities in its planning and implementation. If designed in this way, results from the process study can play a central role in selecting and optimizing a suite of observations that can best meet the broad requirements for more robust, sustained circumpolar and regional observations to meet different demands, including initial conditions for models, improved process understanding, model validation, and long-term prediction.

Influxes of oceanic heat to the Arctic from the Pacific Ocean and the Atlantic Ocean have likely contributed to the recent loss of Arctic sea ice (e.g., Shimada et al, 2006; Walsh et al., 2011). In both regions, the warmer water subducts and circulates below the fresh surface layer of the Arctic Ocean. The rates, locations, and processes by which these heat sources reach the overlying sea ice cover and affect sea ice anomaly evolution are poorly understood, in large part because of a lack of in situ observations. A mechanism involving reductions of ice concentration, increased responsiveness to wind forcing, and enhanced mixing has been proposed by Shimada et al. (2006), while processes of double diffusion and eddy mixing have been suggested as mechanisms by which Atlantic water heat may move upward in the water column north of Eurasia (Polyakov et al., 2011).

Compounding the uncertainty about the role of ocean processes is the varying ability of global and regional ocean models to reproduce the vertical structure of temperature and salinity in the upper layers of the Arctic Ocean. Inadequacies in process formulation as well as vertical resolution are likely sources of errors in the upper Arctic Ocean in global and regional models. Moreover, the CMIP5 decadal predictability experiments, which are targeting seasonal to decadal predictability inherent in ocean initializations, have yet to address the role of the Arctic Ocean in interannual-to-decadal prediction. This is largely because there are few data to initialize the Arctic Ocean in decadal-scale hindcast experiments.

Key Strategy: Model Sensitivity Studies to Determine Key, First-Order Observational Needs

There is a particular need at this time for a coordinated effort to design and implement a set of model sensitivity studies that will provide quantitative metrics to assess the impact of various observation types, locations, and densities on seasonal sea ice forecasts.

Observations play a critical role in seasonal and decadal sea ice prediction. There are broad requirements for more robust sustained circumpolar and regional observations to meet different demands, for instance, initial and boundary conditions for models, improved process understanding, model validation, and long-term prediction. Sea ice model predictive capabilities have evolved and will continue to evolve. That said, to the extent that particular processes (e.g., ocean heat fluxes and cloud-radiative interactions) are found to exert high leverage on the models' simulation of the Arctic system and sea ice prediction in particular, the diagnostic model intercomparison can point to priorities for observations and/or process studies targeted at improved model simulations and predictions of sea ice. This should build on previous work that has identified areas of systematic model bias (e.g., Sorteberg et al., 2007) and the importance of various simulated feedbacks (e.g., Winton, 2006; Kay et al., 2012).

The modeling infrastructure, especially as it pertains to data assimilation, has advanced sufficiently that one can now envision a series of sensitivity studies designed to strategically inform research investments related to observational needs for sea ice modeling and prediction.

At the seasonal timescale, observing system experiments (OSEs—testing impacts of actual observations) and observing system simulation experiments (OSSEs—testing sensitivities to hypothetical or simulated observations) can be directed at systematically investigating the effects of specific observations on prediction capabilities. Among the limited studies to date, Inoue et al. (2009) have shown that the assimilation of sea level pressure measurements from the International Arctic Buoy Program improve atmospheric reanalyses. However, impacts on atmospheric reanalyses are not equivalent to impacts on seasonal sea ice predictions (and sea ice simulations in general). Model-derived predictions of sea ice for timescales of several seasons are almost certainly affected by the initialization of ice thickness (and corresponding distributions of ice concentration), but the atmospheric initialization probably has little effect. There is large uncertainty, however, about the importance of initializations of other variables, such as snow on sea ice, ocean temperature and salinity profiles below the ice, and ocean current distributions.

The modeling tools to conduct these data-model synthesis experiments exist. For instance, coupled atmosphere-ice-ocean forecast models, such as the National Centers for Environmental Prediction's (NCEP's) Coupled Forecast System, routinely assimilate observational data in producing seasonal forecasts of the atmosphere, ocean, and sea ice out to a range of a year. Ice-ocean models have been developed and used to produce forecasts for the SEARCH Sea Ice Outlook[1] (Figure 3.1), as well as for ice-ocean model intercomparisons. In addition, data assimilation systems have been developed for reanalyses of the Arctic atmosphere and ocean (Bromwich et al., 2010; Panteleev et al., 2011).

It is noteworthy that whereas all of the forecasts represented in Figure 3.1 projected a 2012 September minimum ice extent that

---

[1] http://www.arcus.org/search/seaiceoutlook/index.php.

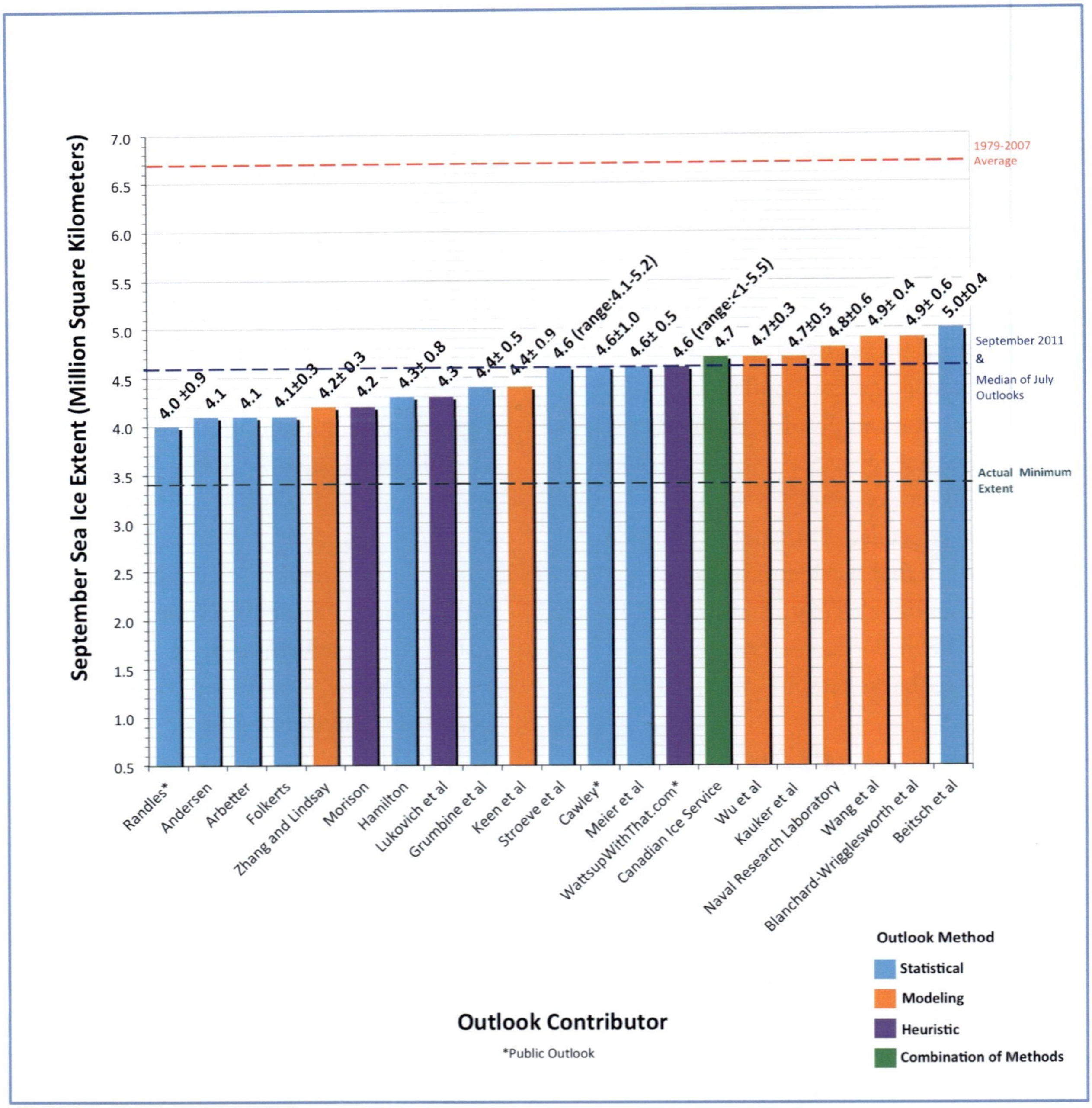

FIGURE 3.1 July 2012 Pan-Arctic Sea Ice Outlook. Ice-ocean models have been developed and are one method used to produce forecasts for the SEARCH Sea Ice Outlook. This figure shows 21 community estimates of the expected minimum of sea ice for 2012. The projected Arctic sea ice extent median value for September 2012 was 4.6 million square kilometers. However, on September 16, 2012, Arctic sea ice reached a minimum extent of 3.41 million square kilometers, the lowest seasonal minimum extent on satellite record. SOURCE: Adapted from ARCUS and SEARCH.

was well below the long-term (1979-2007) average, they all also overestimated it. This result suggests that although there is a skill in the seasonal forecasts relative to climatology (albeit a changing climatology), there remains the need for a concerted effort to improve sea ice predictions—the need that motivated the present report.

In the case of seasonal predictions, the existing capabilities in data-model synthesis have yet to be exploited in the Arctic because prediction capabilities at these timescales are relatively new and because the Arctic observational and modeling communities have tended to be distinct. Therefore, as a first step in designing OSEs and OSSEs, there is a particularly urgent need for a coordinated effort by these communities to design a set of experiments that will provide quantitative metrics of the impact of various observation types, locations, and densities on seasonal sea ice forecasts, as well as the accuracy and temporal resolution that are required.

Possible observational variables for inclusion in these experiments are ice thickness distributions, ice extent and concentration, snow on sea ice, and the upper-ocean profiles of temperature, salinity and current velocities. The latter category of observations includes ocean measurements not only from under the ice, but from the surrounding open ocean. In addition to these state variables, consideration should be given to measurements focused on the exchange of energy between the air-ice and ice-ocean boundaries, which drive ice dynamics and thermodynamics (e.g. radiation, sensible heat, moisture, and momentum) The OSEs and OSSEs would address impacts of measurement errors as well as varying distributions of measurements.

These types of experiments may be regarded as prerequisites for the design of an Arctic observing network (NRC, 2006), and seasonal sea ice prediction can provide a compelling focus for such experiments. Further, many of the variables listed above (e.g., ice thickness, snow on sea ice, and under-ice ocean profiles) are observational challenges in their own right. The logistics and expenses involved in obtaining these measurements adds to the urgency of OSEs and OSSEs to justify, for sea ice prediction and for other applications, future investments in the observations.

Key Strategy: Enhanced Numerical Model Capabilities

Enhancement of model-based predictive capabilities will require coordinated experiments to (a) identify which variables and processes are critical to simulating a realistic ice cover, (b) investigate the source of climate model drift, and (c) guide decisions regarding high-priority model development needs and the expansion of models to include additional capabilities and variables of interest to stakeholders.

Model intercomparison projects (MIPs), such as for the Arctic (AMIP[2]), the Arctic Ocean (AOMIP[3]), and sea ice (SIMIP[4]), have played an important role in identifying

[2]www-pcmdi.llnl.gov/projects/amip/index.php/

[3] www.whoi.edu/page.do?pid=29836

[4] http://gaim.unh.edu/Structure/Future/MIPs/SIMIP.html

BOX 3.4
HURRICANE FORECAST IMPROVEMENT PROGRAM

The NOAA-led Hurricane Forecast Improvement Program (HFIP) is a 10-year program started in 2007 with the overarching goal of improving hurricane forecast skill, with an emphasis on hurricane intensity and structure.[a] Hurricane forecasts have improved significantly in the last 10 years, but rapid intensification remains a significant challenge.

The specific goals of HFIP are to: 1) improve the accuracy of hurricane intensity and track forecasts and 2) increase the forecast confidence of customers and decision makers, especially those in the emergency management community (NOAA, 2010). To help facilitate these goals, the National Oceanic and Atmospheric Association has partnered with other federal agencies, the National Center for Atmospheric Research, universities, and the Naval Research Laboratory (NRL). These collaborations help address the challenges associated with transitioning new forecasting research and technology into operations. Furthermore, there has been an effort within HFIP to ensure open access to the data involved (NOAA, 2010).

[a] www.hfip.org/

differences among models and, in some cases, the component(s) of the model that is the source of those differences. To date, however, MIP efforts have had limited success in pinpointing the variables or processes that are the root cause of simulation errors, and any conclusions have often not efficiently fed back to the model developers to improve the models.

A new strategy for model intercomparison is needed that will identify specific, key processes of importance to sea ice prediction; incorporate lessons learned from model sensitivity studies; and collaborate closely with model developers to identify approaches to resolve unrealistic model behavior. Regional models and ice-ocean coupled systems will likely be an essential part of the strategy, given the greater control achieved in these approaches by prescribed (e.g., observationally- or reanalysis-derived) lateral and/or surface forcing of the Arctic. Interestingly, one outcome of these studies, along with the identification of factors that influence sea ice prediction skill, may be to realize simplifications that can be applied to coupled models. This result would allow for models with reduced complexity to be used for seasonal-scale sea ice prediction.

At the decadal timescale, where predictions are largely influenced by forcing, model sensitivity studies explore and quantify the impact of a range of parameters or representations of physical processes on predicted model outputs. These experiments show how a particular scenario may be affected by multiple parameters. Performing targeted sensitivity studies with new parameterizations can reveal weaknesses of other parameterizations. These simulations make it possible to analyze the sensitivity of simulation results to some of the decisions made in model development. The weather community may be able to contribute lessons learned for conducting sensitivity studies (Box 3.4).

As the Arctic ice cover transitions to a predominantly thinner state, the decadal model sensitivities among variables affecting ice growth, melt, and movement may change relative to those of the past. Targeted sensitivity studies would help identify which variables and processes are critical to simulating a realistic ice cover in the new state, including the mean climatology, simulated variability of and response to changes in external forcing, and where uncertainties inherent in model parameterizations have the largest influence on simulated sea ice processes. Through collaborative efforts with the model development community, this should feed back into improvements in the physics of the models.

A particularly important issue to address via the model sensitivity studies and intercomparison activities is climate model drift from an observationally-based initialized state, which contaminates predictions on seasonal to decadal timescale. This drift results from systematic biases in coupled climate simulations. Improvements in model simulations are required to address this issue. Research on data assimilation methods and alternative methods of model initialization, for example, by using anomaly fields, will need to be considered. Additionally, various mechanisms to "de-drift" predictions need to be assessed in retrospective studies to determine their utility in realizing useful predictive skill.

Related to the call for targeted model sensitivity studies, there is a need for enhanced capabilities in numerical models to provide useful information on key variables of interest. For example, many climate models do not distinguish between land-fast ice and other sea ice, yet the behavior of land-fast ice is of keen interest to a variety of stakeholders. The date of spring breakup is a particularly influential event for coastal infrastructure and operations, but most models lack sufficient resolution and the specific processes that govern evolution of land-fast ice. Many of these requirements call for predictions that have sufficient spatial detail to resolve the highly varying ice characteristics that occur near the coastline. The nature of seasonal sea ice prediction demands accuracy within a few hundred kilometers of shore and within the marginal ice zone. In addition, forcing from tides and ocean waves may play an important role in sea ice evolution on seasonal to decadal timescales. These factors are not typically considered in large-scale numerical models. Model enhancements that incorporate these and other relevant processes would allow for investigations of their role in sea ice prediction and ultimately result in better predictive skill and more useful information for stakeholders.

One way in which model capabilities can be enhanced is by finer resolution. Recent studies have shown that models with higher horizontal and vertical resolution are able to more realistically simulate certain processes in the atmosphere (e.g. Byrkjedal et al., 2008; Girard and Bekcic, 2005) and ocean (e.g., Fieg-Rudiger et al., 2010) Higher resolution information may also be necessary to meet certain stakeholder needs. However simply increasing model resolution is not a panacea, because enhanced computational and storage costs need to be considered in light of the relevant benefits for sea ice prediction. Moreover, model parameterizations that have been developed for coarse resolutions may not be ideal for considerably finer spatial-scales (e.g., Lipscomb et al., 2007; Girard et al., 2009) and may need to be revisited, requiring further model developments. Nevertheless, with computational resources increasing and likely benefits in terms of simulation quality, increased resolution in global and regional models together with regionally refined and adaptive model grids need to be explored in the context of benefits for sea ice predictive capability.

KNOWLEDGE MANAGEMENT

Key Strategy: Improved Information and Data Management

Given the vast amounts of disparate data on Arctic sea ice and the numerous stakeholders who use these data, there is a need for a coordinated and centralized information hub for Arctic datasets that facilitates timely access to observational and modeling results and encourages sustained communication among stakeholders.

No single organization or agency has adequate resources to systematically undertake the task of robust field observations, data synthesis, and environmental modeling. Collaborative efforts and data sharing are therefore essential. Moreover, data continuity is a fundamental imperative so that long-term Arctic sea ice trends can be ascertained and provided to stakeholders for reliable planning. Sharing data also enables researchers to communicate and collaborate more effectively.

Given the vast amounts of disparate data on Arctic sea ice, background information, model results, observational date, etc. can be difficult to find for the numerous stakeholders who use these data, particularly for new users. The committee acknowledges that a more centralized framework could improve information management (Parsons et al., 2011). Although there are numerous data repositories for climate-relevant data, they tend to be scattered and inconsistently cross-linked. Rich measurement datasets are often reduced to their basic parameters with a loss of important information. Also, data transfer and data transformation at data centers add additional layers of complexity and data latency.

In the committee's opinion, the main purpose of a centralized information hub is to serve as a primary launching pad for searches aimed at gaining access to this wide array of information. The intention is not to recreate existing and diverse resources, but to facilitate the ease of their retrieval. A central information hub would unify the various databases, providing a seamless and consistent system for information and data

BOX 3.5
DATA SHARING: TWO CASE STUDIES

Some progress has been made in data management. Case 1: As of 2011 Shell, ConocoPhillips and Statoil have signed an agreement with the National Oceanic and Atmospheric Association to formally share scientific information that the companies have acquired individually and jointly in the Arctic region.[a]

This collaboration leverages the complementary strengths of NOAA's scientific expertise and the industry's significant offshore resources. Scientific datasets for the Arctic region are shared, including weather and ocean observations, biological information, and sea ice and seafloor mapping studies. NOAA's ability to monitor climate change and provide useful products and services that inform energy exploration activities in the Arctic will likely be improved through the sharing of high-quality data. The integration of these data could also provide a greater national capacity to effectively manage and respond to environmental disasters in an area where limited personnel and facilities exist. Data and information are shared with the public through NOAA's existing outlets. Quality control on all data provided is conducted by NOAA before it is incorporated into its products and services.

Case 2: Since 2011, all National Science Foundation proposals must include a supplementary "data management plan," which is subject to peer review. The primary goal of this new data-sharing policy is to "assure that products of research help NSF achieve its mission to promote the progress of science and engineering."[b] The plan should outline the types of data to be produced, the standards to be used, the policies for data access and reuse, and the plans for archiving.

[a] http://www-static.shell.com/static/usa/downloads/alaska/alaska_arcticmou082311.pdf
[b] www.nsf.gov/geo/geo-data-policies/index.jsp

discovery and access. Recent Web standards provide distributed databases that appear uniform and singular to the user. Therefore it is not necessary to create new archives, but rather to leverage existing infrastructure. A key characteristic of the central information hub and the individual components that lie behind it is the timeliness of the available resources. This is particularly critical for applications related to seasonal sea ice predictions, which require real-time data access for model output, in situ observations, satellite and aircraft survey data, etc. The centralized hub could also serve as an integrating resource, providing access to information on the various elements of the Arctic sea ice system (e.g., ocean, atmosphere, and sea ice.)

A separate, but related issue is long-term data storage limitations. In the climate modeling community, the push toward high-resolution and complex models coupled with diverse stakeholder needs has resulted in a rapid and increasing demand for data storage, analysis and distribution (NRC, 2012a). Thus many climate modeling groups are increasingly limited by the cost of long-term data storage.

Consistent data storage protocols need to be adopted that preserve ancillary data and original sample rates along with interoperability standards for data interchange. Indexes need to be based on agreed metadata vocabularies (e.g., Marine Metadata Interoperability Project[5]) for search and reference efficiency. Data density and timely accessibility of information to and from nonfederal sources are some obstacles to be resolved, although there has been slight progress (Box 3.5).

It is hard to dispute the widespread desire for a central information hub and the value that could be derived from it. The major challenge lies in its initiation. There is no shortage of candidate organizations well suited to facilitate the design and implementation of a central information hub, though effective data management requires resources. Because permanence is a critical characteristic, there would be an expectation of long-term and stable support for this important activity. Given the broad and critical nature of these needs, which reaches beyond the issue of sea ice predictability, it may be most appropriately addressed by a high-level, cross-cutting entity.

[5] https://marinemetadata.org/

4

Conclusion

Steady progress has been made in the understanding of Arctic sea ice cover and its role in the Arctic and global systems. In the committee's view increased understanding of the Arctic sea ice cover is linked to steady improvements in our ability to predict sea ice conditions over seasonal to decadal timescales. These gains are marked by important advances in numerical models, instrumentation, methodologies, analysis, data assimilation, and observational techniques. However, recent dramatic change in the ice cover, accompanied by increased demand for access to this heretofore remote region, has created an urgent and escalated level of need for Arctic sea ice predictions to serve a broad stakeholder community. Added pressure comes with the reality of the limited resources (i.e., funding and infrastructure) available to support timely progress and the likelihood that, in the face of continued warming, the Arctic will remain a dynamic environment for the foreseeable future.

This report outlines key challenges and high-priority strategies to facilitate a transformative change in our predictive capability of sea ice conditions on the seasonal to decadal timescale (Box 4.1). Chief among the strategies is a *deliberately integrative approach*, founded on a sustained and coordinated conversation among the user, modeling, and observation communities. In fact, to be successfully implemented, many of the strategies will require the use of an integrative approach. In some ways, the strategies would also serve as a mechanism for sustained conversation and collaboration among the three communities.

This committed approach is necessary to reveal and address key challenges to our fundamental understanding of the Arctic environment and its connection to the global climate system; to balance high-priority user needs against realistic predictive capabilities; to foster coordinated support of this work within the private and public sectors; and to provide guidance in allocation of resources to support the most promising avenues in addressing the most pressing needs. It is an approach that moves beyond the status quo, which relies heavily on a largely disjointed set of research initiatives that often fail to produce a clear set of priorities. Fortunately, there are a number of precedents that exist to inform the design and implementation of this comprehensive communication network. The more daunting challenges are to determine (1) the entity responsible for coordinating and facilitating this exchange and (2) the approach to ensure its sustained support.

Although the report suggests specific advancement strategies related to the models used to formulate seasonal and decadal sea ice prediction, the same level of specificity is not provided for observations (e.g., types and locations or frequencies of observations). Rather, it is the committee's view that systematically identifying obstacles that prevent models from producing more accurate sea ice predictions at seasonal to decadal timescales will aid in (1) directing and prioritizing process studies, (2) designing observing networks, and (3) focusing model development. This perspective takes into account that the modeling infrastructure has advanced sufficiently to support a series of sensitivity studies designed to strategically inform research investments related to key observational needs. In addition, the committee has identified steps that will advance the modeling capabilities that are essential to sea ice prediction over seasonal to decadal timescales. These steps involve sustained interactions with the observational and user communities, reinforcing the importance of integration across the three communities. The extent to which such integration can be achieved will likely determine the rate at which sea ice prediction capabilities advance.

BOX 4.1
Compilation of Key Challenges and Strategies

GAPS IN OUR UNDERSTANDING

- **Overarching Challenges**
  - Treating Sea Ice as Part of a Global System. - A key to advancing our understanding and predictive capabilities is the treatment of the sea ice cover as an integral part of the complex Arctic system which, in turn, is an integral element of the global system. Adding to this complexity is the broad range of human activities that not only influence the Arctic system, but are also influenced by it.
  - Impacts of the Regime Shift of Arctic Sea Ice. - Understanding how the recent regime shift in the Arctic sea ice cover, resulting in a significant reduction in the amount of multiyear ice compared with first-year ice, affects the processes governing the atmosphere-sea ice-ocean system and the models and observations used to study and predict Arctic sea ice dynamics.
  - Identifying Diverse and Emerging Stakeholder Requirements. - As the Arctic is being transformed by globalization and climate change, new stakeholders with additional and more sophisticated requirements are emerging. Clearly defining these diverse needs as they relate to seasonal to decadal sea ice prediction is crucial to inform the future directions of modeling, observations, and overall research.
- **Challenges in Advancing Predictive Capability**
  - Competing Approaches to Seasonal Sea Ice Prediction. - Although limitations in the various approaches used to generate seasonal forecasts are generally acknowledged, there is a lack of quantitative information about their relative strengths and weaknesses.
  - Observational Requirements for Seasonal Sea Ice Prediction. - Seasonal sea ice prediction capability depends on adequate knowledge of initial ice-ocean conditions, even though the specific requirements associated with "adequate knowledge" have yet to be established. This challenge is compounded by the need for a fast turnaround in acquiring and accessing observational data.
  - Projecting Realistic Forcings and Feedbacks for Decadal Sea Ice Predictions. - A key challenge in coupled climate models is the capability to realistically simulate atmospheric and oceanic conditions of relevance to sea ice variability, including the identification of model processes that contribute to unrealistic forcings and feedbacks.

STRATEGIES FOR THE FUTURE

- **Overarching Strategy**
    - A Deliberately Integrated Approach for Sustained and Coordinated Collaboration Among User, Modeling, and Observation Communities. - A deliberately integrated approach is needed to facilitate coordinated and sustained discussions and collaborations among the user, modeling, and observation communities to inform effective research activities and to set realistic expectations for predictions. Such an approach would need to take full advantage of existing infrastructure and draw from comparable efforts in other fields.
- **Strategies to Improve Sea Ice Predictive Capabilities: Seasonal to Decadal Timescales**
    - Evaluation of Existing Seasonal Prediction Methods. - A coordinated and detailed comparison of the different approaches used to generate seasonal sea ice forecasts could establish baseline expectations for predictive skill and identify priority needs, setting the stage for advances in predictive capability.
    - Process-Based Studies Targeted at Increasingly Prevalent First-Year Ice Cover. - Questions surrounding the impact of the trend toward an increasingly seasonal Arctic sea ice cover could be addressed with the development of a highly coordinated and integrated process-based study, analogous to Surface Heat Budget of the Arctic Ocean (SHEBA) project, focused on understanding oceanic, atmospheric, and terrestrial contributions to seasonal sea ice predictions.
    - Model Sensitivity Studies to Determine Key, First-Order Observational Needs. - There is a particular need at this time for a coordinated effort to design and implement a set of model sensitivity studies that will provide quantitative metrics to assess the impact of various observation types, locations, and densities on seasonal sea ice forecasts.
    - Enhanced Numerical Model Capabilities. - Enhancement of model-based predictive capabilities will require coordinated experiments to (a) identify which variables and processes are critical to simulating a realistic ice cover, (b) investigate the source of climate model drift, and (c) guide decisions regarding high-priority model development needs and the expansion of models to include capabilities and additional variables of interest to stakeholders.
- **Knowledge Management**
    - Improved Information and Data Management. - Given the vast amounts of disparate data on Arctic sea ice and the numerous stakeholders who use these data, there is a need for a coordinated and centralized information hub for Arctic datasets that facilitates the timely access to observational and modeling results, and encourages sustained communication among stakeholders.

References

Alexeev, V. A., I. N. Esau, I. V. Polyakov, S. J. Byam, and S. Sorokina. 2011. Vertical structure of recent Arctic warming from observed data and reanalysis products. *Climatic Change* 111(2):215-239.

Allison, I., N. L. Bindoff, R. A. Bindschadler, P. M. Cox, N. de Noblet, M. H. England, J. E. Francis, N. Gruber, A. M. Haywood, D. J. Karoly, G. Kaser, C. Le Quéré, T. M. Lenton, M. E. Mann, B. I. McNeil, A. J. Pitman, S. Rahmstorf, E. Rignot, H. J. Schellnhuber, S. H. Schneider, S. C. Sherwood, R. C. J. Somerville, K. Steffen, E. J. Steig, M. Visbeck, and A. J. Weaver. 2009. *The Copenhagen Diagnosis: Updating the World on the Latest Climate Science.* Sydney, Australia: University of New South Wales, Climate Change Research Centre. Available at http://www.ccrc.unsw.edu.au/Copenhagen/Copenhagen_Diagnosis_LOW.pdf.

AMAP (Arctic Monitoring and Assessment Programme). 2011. *Snow, Water, Ice and Permafrost in the Arctic (SWIPA): Climate Change and the Cryosphere.* Oslo, Norway: AMAP.

Amundrud, T. L., H. Melling, R. G. Ingram, and S. E. Allen. 2006. The effect of structural porosity on the melting of ridge keels in pack ice. *Journal of Geophysical Research* 111:C06004, doi:10.1029/2005JC002895.

Arctic Council. 2009. *Arctic Marine Shipping Assessment 2009.* Project Report. Protection of the Arctic Marine Environment Working Group. Available at http://library.arcticportal.org/1400/1/AMSA_2009_Report_2nd_print.pdf.

Arrigo, K. R., D. K. Perovich, R. S. Pickart, Z. W. Brown, G. L. van Dijken, K. E. Lowry, M. M. Mills, M. A. Palmer, W. M. Balch, F. Bahr, N. R. Bates, C. Benitez-Nelson, B. Bowler, E. Brownlee, J. K. Ehn, K. E. Frey, R. Garley, S. R. Laney, L. Lubelczyk, J. Mathis, A. Matsuoka, B. G. Mitchell, G. W. K. Moore, E. Ortega-Retuerta, S. Pal, C. M. Polashenski, R. A. Reynolds, B. Schieber, H. M. Sosik, M. Stephens, and J. H. Swift. 2012. Massive phytoplankton blooms under Arctic sea ice. *Science* 336(6087):1408.

Arrigo, K. R., and G. L. van Dijken. 2011. Secular trends in Arctic Ocean net primary production. *Journal of Geophysical Research* 116, C09011.

Barber, D. G., M. G. Asplin, R. L. Raddatz , L. M. Candlish, S. Nickels, S. Meakin, K. P. Hochheim, J. V. Lukovich, R. J. Galley, and S. J. Prinsenberg. 2012.

Change and variability in sea ice during the 2007-2008 Canadian International Polar Year program. *Climatic Change* 115(1):115-133.

Barnston, A. G., H. M. van den Dool, D. R. Rodenhuis, C. R. Ropelewski, V. E. Kousky, E. A. O'Lenic, R. E. Livezey, S. E. Zebiak, M. A. Cane, T. P. Barnett, N. E. Graham, M. Ji, and A. Leetmaa. 1994. Long-lead seasonal forecasts—Where do we stand? *Bulletin of the American Meteorological Society* 75:2097-2114.

Barnston, A. G., M. K. Tippett, M. L. L'Heureux, S. Li, and D. G. DeWitt. 2012. Skill of real-time seasonal ENSO model predictions during 2002-11: Is our capability increasing? *Bulletin of the American Meteorological Society* 93:631-651.

Bhatt, U. S., M. A. Alexander, C. Deser, J. E. Walsh, J. S. Miller. M. S, Timlin, J. Scott, and R. A. Tomas, 2008: The atmospheric response to realistic reduced summer sea ice anomalies. *Geophysical Monographs* 180:91-110.

Bhatt, U. S., D. A. Walker, M. K. Raynolds, J. C. Comiso, H. E. Epstein, G. Jia, R. Gens, J. E. Pinzon, C. J. Tucker, C. E. Tweedie, and P. J. Webber. 2010. Circumpolar Arctic Tundra Vegetation Change Is Linked to Sea Ice Decline. *Earth Interactions* 14(8): 1-20.

Black, R. X. 2002. Stratospheric forcing of surface climate in the Arctic Oscillation. *Journal of Climate* 15:268-277.

Blanchard-Wrigglesworth, E., C. M. Bitz, and M. M. Holland. 2011. Influence of initial conditions and climate forcing on predicting Arctic sea ice. *Geophysical Research Letters* 38:L18503, doi:10.1029/2011GL048807.

Bromwich, D., Y.-H. Kuo, M. Serreze, J. Walsh, L.-S. Bai, M. Barlage, K. Hines, and A. Slater. 2010. Arctic system reanalysis: Call for community involvement, *Eos, Transactions, American Geophysical Union* 91(2):13.

Byrkjedal, Ø., I. Esau, and N. G. Kvamstø. 2008. Sensitivity of simulated wintertime Arctic atmosphere to vertical resolution in the ARPEGE/IFS model. *Climate Dynamics* 30:687-701.

Cohen, J., and J. Jones. 2011. Tropospheric precursors and stratospheric warmings. *Journal of Climate* 24:6562-6572.

Comiso, J. 2012. Large decadal decline of the Arctic multiyear ice cover. *Journal of Climate* 25:1176-1193.

Deming, J., and L. Fortier. 2011. Introduction to the special issue on the biology of the circumpolar flaw lead (CFL) in the Amundsen Gulf of the Beaufort Sea (Arctic Ocean). *Polar Biology* 34(12):1797-1801.

Deser, C., R. Tomas, M. Alexander, and D. Lawrence. 2010. The seasonal atmospheric response to projected Arctic sea ice loss in the late 21st century. *Journal of Climate* 23:333-351.

Deser, C., A. Phillips, V. Bourdette, and H. Teng. 2012: Uncertainty in climate change projections: The role of internal variability. *Climate Dynamics* 38:527-546.

Drobot, S. D., J. A. Maslanik, and C. Fowler. 2006. A long-range forecast of Arctic summer sea-ice minimum extent.

Geophysical Research Letters 33:L10501, doi:10.1029/2006GL026216.

Fetterer, F., K. Knowles, W. Meier, and M. Savoie. 2009. Sea Ice Index (digital media). Boulder, CO: National Snow and Ice Data Center.

Fieg-Rudiger, K., E. Fahrbach, A. Beszczynska-Muller, and U. Schauer. 2010. Simulation of oceanic volume transports through Fram Strait 1995-2005. *Ocean Dynamics* 60:491-502.

Francis, J. A., and S. J. Vavrus. 2012. Evidence linking Arctic amplification to extreme weather in mid-latitudes. *Geophysical Research Letters* 39:L06801, doi:10.1029/2012GL051000.

Girard, E., and B. Bekcic. 2005. Sensitivity of an Arctic Regional Climate Model to the horizontal resolution during winter: Implications for aerosol simulation. *International Journal of Climatology* 25:1455-1471.

Girard, L., J. Weiss, J. Molines, B. Barnier, and S. Bouillon. 2009. Evaluation of high-resolution 5 sea ice models on the basis of statistical and scaling properties of Arctic Sea ice drift and deformation. *Journal of Geophysical Research* 114:C08015, doi:10.1029/2008JC005182.

Goosse, H., O. Arzel, C. M. Bitz, A. de Montety, M. Vancoppenolle. 2009. Increased variability of the Arctic summer ice extent in a warmer climate. *Geophysical Research Letters* 36:L23702, doi:10.1029/2009GL040546.

Guemas, V., and D. Salas-Melia. 2008. Simulation of the Atlantic meridional overturning circulation in an atmosphere-ocean global coupled model, I: A mechanism governing the variability of ocean convection in a preindustrial experiment. *Climate Dynamics* 31(1):29-48.

Guilyardi, E., A. Wittenberg, A. Fedorov, M. Collins, C. Wang, A. Capotondi, G. J. van Oldenborgh, and T. Stockdale. 2009. Understanding El Niño in ocean-atmosphere general circulation models: Progress and challenges. *Bulletin of the American Meteorological Society* 90:325-340.

Haas, C., A. Pfaffling, S. Hendricks, L. Rabenstein, J.-L. Etienne, and I. Rigor. 2008. Reduced ice thickness in Arctic Transpolar Drift favors rapid ice retreat. *Geophysical Research Letters* 35:L17501, doi:10.1029/2008GL034457.

Hezel, P. J., X. Zhang, C. M. Bitz, B. P. Kelly, and F. Massonnet. 2012. Projected decline in spring snow depth on Arctic sea ice caused by progressively later autumn open ocean freeze-up this century. *Geophysical Research Letters* 39:L17505, doi:10.1029/2012GL052794.

Hibler, W. D. 1979. A dynamic thermodynamic sea ice model. *Journal of Physical Oceanography* 9:815-846.

Holland, M. M., and J. C. Stroeve. 2011. Changing seasonal sea ice predictor relationships in a changing Arctic climate. *Geophysical Research Letters* 38:L18501, doi:10.1029/2011GL049303.

Holland, M. M., C. M. Bitz, and B. Tremblay. 2006. Future abrupt reductions in the summer Arctic sea ice. *Geophysical Research Letters* 33:L23503, doi:10.1029/2006GL028024.

Holland, M. M., D. A. Bailey, and S. Vavrus. 2011. Inherent sea ice predictability in the rapidly changing Arctic environment of the Community Climate System Model, version 3. *Climate Dynamics* 36:1239-1253.

Hopkins, M. A., and A. S. Thorndike. 2006. Floe formation in Arctic sea ice. *Journal of Geophysical Research* 111:C11S23, doi:10.1029/2005JC003352.

Hutchings, J. K., and I. G. Rigor. 2012. Role of ice dynamics in anomalous ice conditions in the Beaufort Sea during 2006 and 2007. *Journal of Geophysical Research* 117:(C00E04, doi:10.1029/2011JC007182.

Inoue, J., T. Enomoto, T. Miyoshi, and S. Yamane. 2009. Impact of observations from Arctic drifting buoys on the reanalysis of surface fields. *Geophysical Research Letters* 36:L08501, doi:10.1029/2009GL037380.

IPCC (Intergovernmental Panel on Climate Change). 2007. *The Physical Science Basis. Contribution of Working Group I to the Fourth Assessment Report of the IPCC,* S. Solomon, D. Qin, M. Manning, Z. Chen, M. Marquis, K. B. Averty, M. Tignor, and H. L. Miller, eds. Cambridge, UK: Cambridge University Press.

Jackson, J. M., E. C. Carmack, F. A. McLaughlin, S. E. Allen, and R. G. Ingram, 2010: Identification, characterization, and change of the near-surface temperature maximum in the Canada Basin, 1993-2008. *Journal of Geophysical Research* 115:C05021, doi:10.1029/2009JC005265.

Jahn, A., D. A. Bailey, C. M. Bitz, M. M. Holland, E. C. Hunke, J. E. Kay, W. H. Lipscomb, J. A. Maslanik, D. Pollak, K. Sterline, and J. Stroeve. 2012. Late 20th century simulation of Arctic sea ice and ocean properties in CCSM. *Journal of Climate* 25(5):1431-1452. doi:10.1175/JCLI-D-11-00201.1.

Jakobsson, M., L. Mayer, B. Coakley, J. A. Dowdeswell, S. Forbes, B. Fridman, H. Hodnesdal, R. Noormets, R. Pedersen, M. Rebesco, H. W. Schenke, Y. Zarayskaya, D. Accettella, A. Armstrong, R. M. Anderson, P. Bienhoff, A. Camerlenghi, I. Church, M. Edwards, J. V. Gardner, J. K. Hall, B. Hell, O. Hestvik, Y. Kristoffersen, C. Marcussen, R. Mohammad, D. Mosher, S. V. Nghiem, M. T. Pedrosa, P. G. Travaglini, and P. Weatherall. 2012. The International Bathymetry Chart of the Arctic Ocean (IBCAO). *Geophysical Research Letters* 39:L12609, doi:10.1029/2012GL052219.

Kay, J. E., M. M. Holland, and A. Jahn. 2011. Inter-annual to multi-decadal Arctic sea ice extent trends in a warming world. *Geophysical Research Letters* 38:L15708, doi:10.1029/2011GL048008.

Kay, J. E., B. Hillman, S. Klein, Y. Zhang, B. Medeiros, G. Gettelman, R. Pincus, B. Eaton, J. Boyle, R. Marchand, and T. Ackerman. 2012. Exposing global cloud biases in the Community Atmosphere Model (CAM) using satellite observations and their corresponding instrument simulators. *Journal of Climate* 25:5190-5207.

Kelly, B. P., M. Ponce, D. A. Tallmon, B. J. Swanson, and S. K. Sell. 2009. *Genetic Diversity of Ringed Seals Sampled at Breeding Sites; Implications for Population Structure and Sensitivity to Sea Ice Loss.* Final Report 631. Univeristy of Alaska Southeast, North Pacific Research Board. Available at http://doc.nprb.org/web/06_prjs/631_FINAL%20REPORT.pdf.

Kelly, B. P., J. L. Bengtson, P. L. Boveng, M. F. Cameron, S. P. Dahle, J. K. Jansen, E. A. Logerwell, J. E. Overland, C. L. Sabine, G. T. Waring, and J. M. Wilder. 2010a. *Status Review of the Ringed Seal (Phoca hispida).* NOAA Technical Memorandum NMFS-AFSC-212. Washington, DC: U.S. Department of Commerce. Available at http://www.afsc.noaa.gov/Publications/AFSC-TM/NOAA-TM-AFSC-212.pdf.

Kelly, B. P., A. Whiteley, and D. Tallmon. 2010b. The Arctic melting pot. *Nature* 468(7326):891.

Koenigk, T., C. Beatty, M. Caian, R. Doscher, and K. Wyser. 2012. Potential decadal predictability and its sensitivity to sea ice albedo parameterization in a global coupled model. *Climate Dynamics* 38:2389-2408.

Kurtz, N., M. Studinger, S. Farrell, J. Paden, J. Richter-Menge, J. Sonntag, and J. Yungel. 2012. New airborne survey data reveals current state of Arctic sea ice. Submitted to *EOS.*

Lawrence, D. M., A. G. Slater, R. A. Tomas, M. M. Holland, and C. Deser. 2008: Accelerated Arctic land warming and permafrost degradation during rapid sea ice loss. *Geophysical Research Letters* 35:L11506, doi:10.1029/2008GL033985.

Lindsay, R. W., J. Zhang, A. Schweiger, M. Steele, and H. Stern. 2009. Arctic sea ice retreat in 2007 follows thinning trend. *Journal of Climate* 22:165-176.

Lipscomb, W., E. Hunke, W. Maslowski, and J. Jakacki. 2007. Ridging, strength, and stability in high-resolution sea ice models. *Journal of Geophysical Research* 112:C03S91, doi:10.1029/2005JC003355.

Liu, J., J. A. Curry, H. Wang, M. Song, and R. M. Horton. 2012. Impact of declining Arctic sea ice on winter snowfall. *Proceeding of the National Academy of Sciences of the United States of America* 109(11):4074-4079.

Mahoney, A., H. Eicken, A. Gaylord, and L. Shapiro. 2007. Alaska landfast sea ice: Links with bathymetry and atmospheric circulation. *Journal of Geophysical Research* 112:C02001, doi:10.1029/2006JC003559.

Maslanik, J., J. Stroeve, C. Fowler, and W. Emery. 2011. Distribution and trends in Arctic sea ice age through spring 2011. *Geophysical Research Letters* 38(13):L13502, doi:10.1029/2011GL047735.

Maslowski, W., J. Kinney, M. Higgins, and A. Roberts. 2012. The future of Arctic sea ice. *Annual Review of Earth and Planetary Sciences* 40:625-654.

Massonnet, F., T. Fichefet, H. Goosse, C. M. Bitz, G. Philippon-Berthier, M. M. Holland, P.-Y. Barriat. 2012. The trends in summer Arctic sea ice extent are nonlinearly related to the mean sea ice state in CMIP5 models. *Geophysical*

Research Letters, submitted. Available at http://www.uib.no/People/ngfhd/EarthClim/Publications/Papers/massonnet_etal_2012.pdf.

McPhaden, M. J., S. E. Zebiak, and M. H. Glantz. 2006. ENSO as an integrating concept in earth science. *Science* 314:1739-1745.

Meehl, G. A., T. F. Stocker, W. D. Collins, P. Friedlingstein, A. T. Gaye, J. M. Gregory, A. Kitoh, R. Knutti, J. M. Murphy, A. Noda, S. C. B. Raper, I. G. Watterson, A. J. Weaver and Z.-C. Zhao. 2007. Global climate projections. In *Climate Change 2007: The Physical Science Basis. Contribution of Working Group I to the Fourth Assessment Report of the Intergovernmental Panel on Climate Change*, S. Solomon, D. Qin, M. Manning, Z. Chen, M. Marquis, K. B. Averyt, M. Tignor, and H.L. Miller, eds. Cambridge, UK, and New York: Cambridge University Press.

Melling, H., D. A. Riedel, and Z. Gedalof. 2005. Trends in the draft and extent of seasonal pack ice, Canadian Beaufort Sea. *Geophysical Research Letters* 32:L24501, doi:10.1029/2005GL024483.

Min, S.-K., X. Zhang, F. W. Zwiers, and T. Agnew. 2008. Human influence on Arctic sea ice detectable from early 1990s onwards. *Geophysical Research Letters* 35:L21701, doi:10.1029/2008GL035725.

Moss, R. H., J. A. Edmonds, K. A. Hibbard, M. R. Manning, S. K. Rose, D. P. van Vuuren, T. R. Carter, S. Emori, M. Kainuma, T. Kram, G. A. Meehl, J. F. B. Mitchell, N. Nakicenovic, K. Riahi, S. J. Smith, R. J. Stouffer, A. M. Thomson, J. P. Weyant, and T. J. Wilbanks. 2010. The next generation of scenarios for climate change research and assessment. *Nature* 463:747-756. Available at http://dx.doi.org/10.1038/nature08823.

NRC (National Research Council). 2006. *Toward an Integrated Arctic Observing Network*. Washington, DC: The National Academies Press.

NRC. 2010. *Advancing the Science of Climate Change*. Washington, DC: The National Academies Press.

NRC. 2012a. *A National Strategy for Advancing Climate Modeling*. Washington, DC: The National Academies Press.

NRC. 2012b. *Lessons and Legacies of the International Polar Year 2007-2008*. Washington, DC: The National Academies Press.

National Snow & Ice Data Center (NSIDC) Cryosphere Glossary. Accessed July 31, 2012. http://nsidc.org/cgi-bin/words/glossary.pl.

Nghiem, S. V., G. Rigor, D. K. Perovich, P. Clemente-Colón, J. W. Weatherly, and G. Neumann. 2007. Rapid reduction of Arctic perennial sea ice. *Geophysical Research Letters* 34:L19504, doi:10.1029/2007GL031138.

NOAA (National Oceanic and Atmospheric Administration). *Status Review of the Bearded Seal (Erignathus barbatus)*. NOAA Technical Memorandum NMFS-AFSC-211-2010. Available at http://www.afsc.noaa.gov/techmemos/nmfs-afsc-211.htm. Accessed July 31, 2012.

Notz, D., and J. Marotzke. 2012. Observations reveal external driver for Arctic sea-ice retreat. *Geophysical Research Letters* 39:L08502, doi:10.1029/2012GL051094.

Ogi, M., K. Yamazaki, and J. M. Wallace. 2010. Influence of winter and summer surface wind anomalies on summer Arctic sea ice extent. *Geophysical Research Letters* 37:L07701, doi:10.1029/2009GL042356.

Ogi, M. and J. M. Wallace. 2012. The role of summer surface wind anomalies in the summer Arctic sea ice extent in 2010 and 2011. *Geophysical Research Letters* 39(L09704).

Overland, J. E., and M. Wang. 2010. Large-scale atmospheric circulation changes are associated with the recent loss of Arctic sea ice. *Tellus* 62A:1-9.

Overland, J. E. E., J. Francis, E. Hanna, and M. Wang. 2012. The recent shift in early summer Arctic atmospheric circulation. *Geophysical Research Letters* 39:L19804, doi:10.1029/2012GL053268.

Panteleev, G. G., M. Yaremchuk, P. J. Stabeno, V. Luchin, D. A. Nechaev, and T. Kikuchi. 2011. Dynamic topography of the Bering Sea. *Journal of Geophysical Research Oceans* 116:C05017, doi:10.1029/2010JC006354.

Parsons, M. A., Ø. Godøy, E. LeDrew, T. F. de Bruin, B. Danis, S. Tomlinson, and D. Carlson. 2011. A conceptual framework for managing very diverse data for complex, interdisciplinary science. *Journal of Information Science* 37(6):555-569.

Perovich, D. K., and C. Polashenski. 2012. Albedo evolution of seasonal Arctic sea ice. *Geophysical Research Letters* 39:L08501, doi:10.1029/2012GL051432.

Perovich, D. K., E. L. Andreas, J. A. Curry, H. Eiken, C. W. Fairall, T. C. Grenfell, P. S. Guest, J. Intrieri, D. Kadko, R. W. Lindsay, M. G. McPhee, J. Morison, R. E. Moritz, C. A. Paulson, W. S. Pegau, P. O. G. Persson, R. Pinkel, J. A. Richter-Menge, T. Stanton, H. Stern, M. Sturm, W. B. Tucker III, and T. Uttal. 1999. Year on the ice gives climate insights. *Eos, Transactions, American Geophysical Union* 80(481):485-486.

Perovich, D. K., R. C. Moritz, and J. Weatherly. 2003. SHEBA: The Surface Heat Budget of the Arctic Ocean. *Arctic Research of the United States* 17(03048):18-23.

Perovich, D. K., B. Light, H. Eicken, K. F. Jones, K. Runciman, and S. V. Nghiem. 2007. Increasing solar heating of the Arctic Ocean and adjacent seas, 1979-2005: Attribution and role in the ice-albedo feedback. *Geophysical Research Letters* 34:L19505, doi:10.1029/2007GL031480.

Perovich, D. K., J. A. Richter-Menge, K. F. Jones, and B. Light. 2008. Sunlight, water, and ice: Extreme Arctic sea ice melt during the summer of 2007. *Geophysical Research Letters* 35:L1150, doi:10.1029/2008GL034007.

Perovich, D., W. Meier, J. Maslanik, and J. Richter-Menge. 2011. Sea ice. In *Arctic Report Card 2011.* Available at http://www.arctic.noaa.gov/reportcard/sea_ice.html.

Pohlmann, H., J. H. Jungclaus, A. Köhl, D. Stammer, and J. Marotzke, 2009. Initializing decadal climate predictions with the GECCO oceanic synthesis: Effects on the North Atlantic. *Journal of Climate* 22:3926-3938.

Polashenski, C., D. Perovich, J. Richter-Menge, and B. Elder. 2011. Seasonal ice mass balance buoys: Adapting tools to the changing Arctic. *Annals of Glaciology* 52(57):18-26.

Polyakov, I. V, V. A. Alexeev, U. S. Bhatt, E. I. Polyakova, and X. Zhang. 2010. North Atlantic warming: Fingerprints of climate change and multidecadal variability. *Climate Dynamics.* 34:439-457

Polyakov, I. V., V. A. Alexeev, I. M. Ashik, S. Bacon, A. Beszczynska-Möller, E. C. Carmack, I. A. Dmitrenko, L. Fortier, J.-C. Gascard, E. Hansen, J. Hölemann, V. V. Ivanov, T. Kikuchi, S. Kirillov, Y.-D. Lenn, F. A. McLaughlin, J. Piechura, I. Repina, L. A. Timokhov, W. Walczowski, and R. Woodgate. 2011. Fate of early-2000's Arctic warm water pulse. *Bulletin of the American Meteorological Society* 92:561-566..

Rampal, P., J. Weiss, D. Marsan, and M. Bourgoin. 2009. Arctic sea ice velocity field: General circulation and turbulent-like fluctuations. *Journal of Geophysical Research* 114:C10014, doi:10.1029/2008JC005227.

Randall, D. A., R.A. Wood, S. Bony, R. Colman, T. Fichefet, J. Fyfe, V. Kattsov, A. Pitman, J. Shukla, J. Srinivasan, R. J. Stouffer, A. Sumi, and K. E. Taylor. 2007. Climate models and their evaluation. In *Climate Change 2007: The Physical Science Basis, Contribution of Working Group I to the Fourth Assessment Report of the Intergovernmental Panel on Climate Change,* S. Solomon, D. Qin, M. Manning, Z. Chen, M. Marquis, K. B. Averyt, M. Tignor, and H. L. Miller, eds. Cambridge, UK, and New York: Cambridge University Press, pp. 589-662.

Ray, G. C., J. E. Overland, and G. L. Hufford. 2010. Seascape as an organizing principle for evaluating walrus and seal sea-ice habitat in Beringia. *Geophysical Research Letters* 37(20):L20504, doi: 10.1029/2010GL044452.

Rigor, I. G., J. M. Wallace, and R. L. Colony. 2002. Response of sea ice to the Arctic Oscillation. *Journal of Climate* 15:2648-2663.

Shimada, K., T. Kamoshida, M. Itoh, S. Nishino, E. Carmack, F. McLaughlin, S. Zimmermann, and A. Proshutinsky. 2006. Pacific Ocean inflow: Influence on catastrophic reduction of sea ice cover in the Arctic Ocean. *Geophysical Research Letters* 33:L08605, doi:10.1029/2005GL025624.

Simpson, W. R., R. von Glasow, K. Riedel, P. Anderson, P. Ariya, J. Bottenheim, J. Burrows, L. J. Carpenter, U. Frieß, M. E. Goodsite, D. Heard, M. Hutterli, H.-W. Jacobi, L. Kaleschke, B. Neff, J. Plane, U. Platt, A. Richter, H. Roscoe, R. Sander, P. Shepson, J. Sodeau, A. Steffen, T. Wagner, and E. Wolff. 2007. Halogens and their role in polar boundary-layer ozone depletion. *Atmospheric Chemistry and Physics* 7:4375-4418.

Sorteberg, A., V. Kattsov, J. E. Walsh, and T. Pavlova. 2007. The Arctic surface energy budget as simulated with the IPCC AR4 AOGCMs. *Climate Dynamics* 29:131-156.

Steele, M., W. S. Ermold, and J. Zhang, 2011: Modeling the formation and fate of the near-surface temperature maximum in the Canadian Basin of the Arctic Ocean. *Journal of Geophysical Research* 116:C11015, doi:10.1029/2010JC006803.

Stenchikov, G., K. Hamilton, R. J. Stouffer, A. Robock, V. Ramaswamy, B. Santer, and H.-F. Graf. 2006. Arctic Oscillation response to volcanic eruptions in the IPCC AR4 climate model. *Journal of Geophysical Research* 111:D07107, doi:10.1029/2005JD006286.

Stroeve, J. C., M. C. Serreze, M. M. Holland, J. E. Kay, J. Maslanik, and A. P. Barrett. 2011. The Arctic's rapidly shrinking ice cover: A research synthesis. *Climatic Change* 110(3):1005-1027.

Stroeve, J. C., V. Kattsov, A. Barrett, M. Serreze, T. Pavlova, M. Holland, and W. N. Meier. 2012. Trends in Arctic sea ice extent from CMIP5, CMIP3 and observations. *Geophysical Research Letters* 39:L16502, doi:10.1029/2012GL052676.

Strong, C. 2012. Atmospheric influence on Arctic marginal ice zone position and width in the Atlantic sector, February-April 1979-2010. *Climate Dynamics*, in press, doi:10.1007/s00382-012-1356-6.

Sulsky, D., H. Schreyer, K. Peterson, R. Kwok, and M. Coon. 2007. Using the material-point method to model sea ice dynamics. *Journal of Geophysical Research* 112:C02S90, doi:10.1029/2005JC003329

Thomas, P. O., and K. L. Laidre. 2011. Biodiversity—cetaceans and pinnipeds (whales and seals). In *Arctic Report Card 2011*. Available at http://www.arctic.noaa.gov/reportcard.

Thorndike, A. S., D. A. Rothrock, G. A. Maykut, and R. Colony. 1975. The thickness distribution of sea ice. *Journal of Geophysical Research* 80:4501-4513.

Timco, G. W., and W.,F. Weeks. 2010. A review of the engineering properties of sea ice. *Cold Regions Science and Technology* 60(2010):107-129.

Walsh, J. E., J. E. Overland, P.Y. Groisman, and B. Rudolf. 2011. Ongoing climate change in the Arctic. *AMBIO: A Journal of the Human Environment* 40(sp1):6-16.

Wang, J., J. L. Zhang, E. Watanabe, M. Ikeda, K. Mizobata, J. E. Walsh, X. Z. Bai, and B.Y. Wu. 2009. Is the Dipole Anomaly a major driver to record lows in Arctic summer sea ice extent? *Geophysical Research Letters* 36:L05706, doi:10.1029/2008GL036706.

Wang, M., and J. E. Overland. 2012. A sea ice free summer Arctic within 30 years—an update from CMIP5 models. *Geophysical Research Letters*, doi:10.1029/2012GL052868.

Weeks, W. F., and S. F. Ackley. 1982. *The Growth, Structure, and Properties of Sea Ice*. CRREL Monograph 82-1. Hanover, NH: U.S. Army Cold Regions Research and Engineering Laboratory.

Winton, M. 2006. Amplified Arctic climate change: What does surface albedo feedback have to do with it? *Geophysical*

Research Letters 33:L03701, doi:10.1029/2005GL025244.

Wu, B., J. E. Overland, and R. D'Arrigo, 2012. Anomalous Arctic surface wind patterns and their impacts on September sea ice minima and trend. *Tellus* 64:18590, doi:10.3402/tellusa.v64i0.18590.

Yeager, S., A. Karspeck, G. Danabasoglu, J. Tribbia, and H. Teng, 2012. A decadal prediction case study: Late twentieth-century North Atlantic Ocean heat content. *Journal of Climate* 25:5173-5189.

Zhang, J., R. Lindsay, M. Steele, and A. Schweiger. 2008a. What drove the dramatic retreat of Arctic sea ice during summer 2007? *Geophysical Research Letters* 35:L11505, doi:10.1029/2008GL034005.

Zhang, J., M. Steele, R. Lindsay, A. Schweiger, and J. Morison. 2008b. Ensemble 1-year predictions of Arctic sea ice for the spring and summer of 2008. *Geophysical Research Letters* 35:L08502, doi:10.1029/2008GL033244.

APPENDIXES

A
Summary of Recent and Evolving Arctic Sea Ice Predictability Efforts

(Provided to workshop participants as background information)

Forecasts of the sea ice minimum in summer have been collected and synthesized in the Study for Environmental Arctic Change (SEARCH) Sea Ice Outlook (ARCUS, 2010) since 2008, with several research groups in the United States, Canada, and Europe participating. The methods used by these groups vary, with some groups using models, some statistical methods, and others deterministic methods. Overall the goal of the SEARCH Sea Ice Outlook is not to issue sea ice predictions themselves, but to summarize and synthesize available information from the scientific community on the expected September Arctic sea ice minimum.

A related activity is the Arctic Observing Coordination Workshop, which was held in March 2012. The workshop was organized around the SEARCH 5-Year Science Goals and Objectives focusing on sea ice, permafrost, land-ice, and society/policy. The workshop participants concluded that the use of ocean observations to improve sea ice forecasting on various timescales (daily, seasonal, interannual, and decadal) would lead to safe marine operations, infrastructure/community planning, and ecosystem stewardship in the Arctic.

The increased user-demand for sea ice predictions coupled with the current lack of operational sea ice forecasting capability have also sparked two National Oceanic and Atmospheric Administration (NOAA) workshops in 2010 and 2011 (NOAA, 2011), which were tasked with identifying actions NOAA could take over the next few years (2012-2014) to improve its sea ice forecasting capability. The report recommended that the Sea Ice Outlook should not only continue, but should also be converted to a formal program, potentially adding fall freeze-up dates and more detailed regional forecasts to the current predictions of the Arctic-wide sea ice minimum. Further improvements of sea ice models and coupled simulations are also needed to investigate the predictability of sea ice on decadal timescales.

The Arctic Monitoring and Assessment Program (AMAP) also recommended several actions related to Arctic sea ice in its "SWIPA 2011 Executive Summary" (AMAP, 2011) including: maintaining and supporting development of remote sensing methods for observing the cryosphere; expanding research into processes that are important for modeling the cryosphere; and making accurate forecasts for ice, weather, and sea conditions accessible to all Arctic residents and organizations.

Several organizations have focused their efforts on marine operations in the Arctic. An Arctic Roadmap report (US Navy, 2009) from the US Navy Task Force Climate Change outlined a 5-year plan for Navy operations and research in the Arctic. The report recommended the identification of a high-confidence timeline for increased access to the Arctic. It also recommended that the potential for developing high-resolution coupled, air-ocean-ice, prediction capability for the Arctic region be evaluated.

The Arctic Council's Arctic Marine Shipping Assessment 2009 Report (Arctic Council, 2009) focused on current and future marine activity in the Arctic Ocean. The report called for research to improve regional models for increased understanding and enhanced forecasting of regional Arctic sea ice variability. Also noted is the need for comprehensive analyses of current and future global climate model simulations of Arctic sea ice extent to quantitatively assess the range of plausibility of ice-free and partially ice-covered conditions. The report also noted the importance of continued research on Arctic sea ice thickness atmosphere ocean ice forecasting. The Assessment also noted the importance of enhanced ice forecasting and prediction to improving Arctic marine safety and environmental protection (Arctic Council, 2009).

Several reports discuss the need for improved communications and increased stakeholder-involvement. A 2010 workshop by the SEARCH "Understanding Arctic Change Task Force" (ARCUS, 2010) concluded that there must be a clear understanding in the scientific community of what planners and decision makers require for predictions to be useful, and the scientific community must communicate the current predictive capabilities in clear and useful ways to stakeholders, while also quantifying and explaining uncertainties related to them.

Many of the recommendations from a 2010 report from the US Arctic Research Commission (USARC, 2010) focused on stakeholder needs for sea ice forecasts. The report notes that there is a need for communication between scientists, operational forecasting centers, and stakeholders. The decision-making community needs to clearly articulate the space and time domains over which it needs actionable scientific information and the science community needs to assess its readiness to provide this knowledge. These various communities should also hold forums on the issue of uncertainty and how to interpret and use these estimates in a proactive and positive manner.

Some efforts at increasing communication between different scientific communities have already been made. Observational and modeling communities met to discuss future needs for sea ice research at the Climate and Cryosphere (CliC) workshop in November 2011. These discussions sparked the creation of two white papers, outlining the observational needs for advancing sea ice modeling (Massonnet and Jahn, 2012) and polar climate modeling (Kay et al., 2012) as well as highlighting some of the challenges of comparing models with observations.

The World Climate Research Program (WCRP) workshops on sea ice predictability, which occurred in 2010 (WCRP, 2010) and in April 2012, brought together scientists from the modeling and observational sea ice communities. The aim of the workshops was the development of a draft implementation plan for a WCRP polar climate predictability initiative (WCRP, 2012).

REFERENCES

AMAP (Arctic Monitoring and Assessment Programme). Snow, Water, Ice and Permafrost in the Arctic (SWIPA) Executive Summary 2011. Available at http://amap.no/swipa/SWIPA2011ExecutiveSummaryV2.pdf, accessed June 5, 2012.

Arctic Council. 2009. Arctic Marine Shipping Assessment 2009. Project Report. Protection of the Arctic Marine Environment Working Group. Available at http://library.arcticportal.org/1400/1/AMSA_2009_Report_2nd_print.pdf.

ARCUS (Arctic Research Consortium of the United States). 2010. Sea Ice Outlook. Available at http://www.arcus.org/search/seaiceoutlook/index.php, accessed November 8, 2012.

ARCUS. Understanding Arctic Change Workshop, September 29-October 1, 2010. Available at http://www.arcus.org/search/meetings/2010/understanding-arctic-change. Accessed June 5, 2012.

Kay, J. E., G. de Boer, and E. Hunke. 2012. On the observational needs for climate models in polar regions. Available at http://www.cgd.ucar.edu/staff/jenkay/pcwg/PCWG_workingdoc_obs4modelers_march13,2012.pdf, accessed June 5, 2012.

Massonnet, F., and A. Jahn. 2012. Observational needs for sea ice models. Available at http://www.astr.ucl.ac.be/users/fmasson/obs_CLIC_note.pdf, accessed June 5, 2012.

NOAA (National Oceanic and Atmospheric Association). *NOAA Sea Ice Forecasting—Workshop Summary.* Available at http://www.arctic.noaa.gov/docs/NOAA_Sea_Ice_Forecasting_Workshop_Summary.pdf, accessed June 5, 2012.

University of Alaska Fairbanks—University of the Arctic Institute for Applied Circumpolar Policy. 2009. *Considering a Roadmap Forward: The Arctic Marine Shipping Assessment: Workshop Report*, L. W. Brigham and M. P. Sfraga, eds. Available at http://www.uarctic.org/AMSA_workshop_report_final_09.2010_-3FYy.pdf.file, accessed June 5, 2012.

USARC (U.S. Arctic Research Commission). 2010. *Scaling Studies in Arctic System Science and Policy Support: A Call-to-Research*, C. J. Vorosmarty, A. D. McGuire, and J. E. Hobbie, eds. Available at http://www.arctic.gov/publications/arctic_scaling.pdf, accessed June 5, 2012.

U.S. Navy. 2009. U.S. Navy Arctic Roadmap. Available at http://www.navy.mil/navydata/documents/USN_artic_roadmap.pdf, accessed June 5, 2011.

WCRP (World Climate Research Programme). 2010. *WCRP Workshop on*

Seasonal to Multi-Decadal Predictability of Polar Climate, October 25-29, 2010, Bergen, Norway. T.G. Shepherd, J. M. Arblaster, C. M. Bitz, T. Furevik, H. Goosse,V. M. Kattsov, J. Marshall, V. Ryabinin, and J. E. Walsh, eds. Available at http://www.wcrp-climate.org/documents/Polar_WCRP_Report.pdf, accessed June 5, 2012.

WCRP. 2012. *WCRP/IASC Polar Climate Predictability Workshop, 2-4 April 2012, Toronto, Canada.* Available at http://www.atmosp.physics.utoronto.ca/C-SPARC/Polar-WS-website/Polar-Workshop.html. Last modified February 28, 2012.

B
Workshop Information

The committee developed the workshop agenda and invited leading sea ice scientists, experts, and stakeholders to identify obstacles impeding progress in the prediction of Arctic sea ice on seasonal to decadal timescales, and to explore strategies to mitigate those obstacles. To address its task, the committee developed several fundamental working guidelines. The committee considered ice conditions during all seasons within the whole Arctic marine environment (i.e., Arctic Ocean and the subpolar seas, including the seasonal sea ice zone). The committee also provided the participants with a background document that summarized insights and information gained from previously related efforts and published works (see Appendix A). Challenges and strategies were identified during this workshop through presentations, breakout group discussions, and plenary summaries.

Workshop Agenda
May 9-10, 2012
University of Colorado Boulder, East Campus
Administrative and Research Center (ARC)
Boulder, CO

Workshop Goals: Arctic sea ice plays a number of important roles in moderating global climate and influencing oceanic and atmospheric circulation. Recent observed changes in the characteristics of the sea ice cover have various direct and indirect scientific, technological, and societal impacts such as the planning of new shipping ports, oil and gas exploration, increased marine transportation, as well as local and global climate and ecological changes. Currently, our limited understanding of the coupled and complex interactions between Arctic sea ice, oceans, and atmosphere hinders our ability to predict the rate and magnitude of future change. Enhancements of our theoretical, observing, and modeling capabilities will be essential for advancing the understanding and prediction of sea ice over seasonal to decadal timescales. The goal of the workshop is to foster a dialogue between polar scientists, agency representatives, and stakeholders to explore the current major challenges, with a focus on whether there are new methods, observations, and technologies that might advance our predictive capabilities through improved understanding of seasonal to decadal sea ice variations. This dialogue will provide expert information for the preparation of a National Research Council report.

Overarching Questions: What is limiting advancements in sea ice predictions on seasonal to decadal timescales? How can these limitations be overcome to realize necessary advancements?

Wednesday, May 9, 2012

A shuttle will pick up workshop participants from the Boulder Marriott on Canyon Blvd at 7:50 A.M., though participants may walk if they wish.

Room: ARC 620

8:00 A.M. *Breakfast*

8:30 A.M. WELCOME AND INTRODUCTION — Jackie Richter-Menge & John Walsh, *Cochairs*
Purpose of the Study and the Workshop

9:00 A.M. STAKEHOLDER PANEL — Lawson Brigham, *Moderator*
Gary Hufford, NOAA/NWS
Vera Metcalf, Eskimo Walrus Commission
LCDR Kenneth Boda, US Coast Guard
Michael Terminel, Edison Chouest Offshore

Key questions for the panelists:
- What are the key questions you need answers to (and on what timescales?)
- What information, beyond what is currently available to you, do you need to help make decisions?
- What information are you receiving now that is useful to you?

9:35 A.M. DISCUSSION

10:15 A.M. *Break*

SESSION 1 - OBSERVATIONS

10:30 A.M. OBSERVATIONS PANEL — Rebecca Woodgate, *Moderator*
Hajo Eicken, UAF
Walt Meier, NSIDC
Ron Lindsay, UW
Key questions for the panelists:
- What are the key gaps in our understanding?
- What are the key observational challenges in the next five years?
- What advances in observations could address these issues?
- What interactions with modelers and stakeholders would benefit these goals?

10:55 A.M. DISCUSSION

11:30 A.M. BREAKOUTS
Questions for breakout group discussion:
- What are the key challenges and questions?
- What are strategies for addressing these challenges?
- What are the next steps that should be taken?

Blue Group	Green Group
Leader: Jennifer Francis Staff: Katie Thomas Rapporteur: Don Perovich Room # ARC446	Leader: Robert Raye Staff: Amanda Purcell/Chris Elfring Rapporteur: Ignatius Rigor Room # ARC248
Red Group	**Black Group**
Leader: Son Nghiem Staff: Deb Glickson Rapporteur: Ron Kwok Room # RL233	Leader: Jackie Richter-Menge Staff: Lauren Brown Rapporteur: Jim Maslanik Room # RL269

12:30 P.M. *Lunch*

1:45 P.M. REPORT BACK

2:45 P.M. *Break*

SESSION 2 - MODELING

3:15 P.M. MODELING PANEL — Marika Holland, *Moderator*

Cecilia Bitz, UW
Elizabeth Hunke, LANL
Andrey Proshutinsky, WHOI

Key questions for the panelists:

- What are the key gaps in our understanding?
- What are the key modeling challenges in the next five years?
- What advances in modeling could address these issues?
- What interactions with observationalists and stakeholders would benefit these goals?

3:40 P.M. DISCUSSION

4:15 P.M. BREAKOUTS

Questions for breakout group discussion:

- What are the key challenges and questions?
- What are strategies for addressing these challenges?
- What are the next steps that should be taken?

Blue Group	Green Group
Leader: Jennifer Francis Staff: Katie Thomas	Leader: Robert Raye Staff: Amanda Purcell/Chris Elfring

Rapporteur: Sinead Farrell Room # ARC446	Rapporteur: Alex Jahn Room # ARC248
Red Group Leader: Son Nghiem Staff: Deb Glickson Rapporteur: Wieslaw Maslowski Room # RL233	**Black Group** Leader: Jackie Richter-Menge Staff: Lauren Brown Rapporteur: Jenny Hutchings Room # RL269

5:30 P.M. *Adjourn*

Shuttle will be available to take participants back to the hotel.

Thursday, May 10, 2012

Shuttle will pick up workshop participants from the Boulder Marriott on Canyon Blvd at 7:50 A.M., though participants may walk if they wish.

Room: ARC 620

8:00 A.M. *Breakfast*

8:30 A.M. REPORT BACK

SESSION 3—CHALLENGES AND OPPORTUNITIES

9:30 A.M. CHALLENGES AND OPPORTUNITIES PANEL — John Walsh, *Moderator*

Jim Overland, NOAA/PMEL
Brendan Kelly, IARPC
Pablo Clemente-Colón, NOAA/National Ice Center
Jean-Claude Gascard, Université Pierre et Marie Curie

Key questions for the panelists:

- What gaps or questions in sea ice prediction have not yet been addressed in previous efforts or reports?
- What are some cross-cutting issues with observation and modeling interactions?
- How can the various communities (observationalists, modelers, stakeholders) better coordinate?

10:05 A.M. DISCUSSION

10:30 A.M. *Break*

11:00 A.M. BREAKOUTS

Questions for breakout group discussion:

- What are the key challenges and questions?
- What are strategies for addressing these challenges?
- What are the next steps that should be taken?

Blue Group	**Green Group**
Leader: Jennifer Francis Staff: Katie Thomas Rapporteur: Molly McCammon Room # ARC446	Leader: Robert Raye Staff: Amanda Purcell/Chris Elfring Rapporteur: Peter Wadhams Room # ARC248
Red Group	**Black Group**
Leader: Son Nghiem Staff: Deb Glickson Rapporteur: Justin Wettstein Room # RL233	Leader: Jackie Richter-Menge Staff: Lauren Brown Rapporteur: Ed Blanchard-Wrigglesworth Room # RL269

12:15 P.M. *Lunch*

1:15 P.M. REPORT BACK

2:15 P.M. WRAP-UP AND FINAL REMARKS — Jackie Richter-Menge & John Walsh, *Cochairs*

3:00 P.M. *Adjourn*

Shuttle will be available to take participants back to the hotel.

The Future of Arctic Sea Ice Research in Support of Seasonal to Decadal Prediction Participant List

Cecilia Bitz, University of Washington
Ed Blanchard-Wrigglesworth, University of Washington
LCDR Ken Boda, U.S. Coast Guard
Lawson Brigham, University of Alaska
Lauren Brown, National Research Council
Pablo Clemente-Colon, NOAA National Ice Center
Hajo Eicken, University of Alaska Fairbanks
Chris Elfring, National Research Council
John Farrell, U.S. Arctic Research Commission

Sinead Farrell, University of Maryland
Jennifer Francis, Rutgers University
Jean-Claude Gascard, Université Pierre et Marie Curie
Deb Glickson, National Research Council
Jeff Gossett, USN Arctic Submarine Laboratory
Marika Holland, National Center for Atmospheric Research
Amy Holman, NOAA National Ocean Service
Gary Hufford, National Weather Service—Alaska Region
Elizabeth Hunke, Los Alamos National Laboratory
Jenny Hutchings, University of Alaska, Fairbanks
Janet Intrieri, NOAA Earth System Research Laboratory
Alexandra Jahn, National Center for Atmospheric Research
Brendan Kelly, Interagency Arctic Research Policy Committee
Ron Kwok, NASA Jet Propulsion Laboratory
Ron Lindsay, University of Washington
Jim Maslanik, University of Colorado, Boulder
Wieslaw Maslowski, Naval Postgrad School
Larry Mayer, University of New Hampshire
LCDR Blake McBride, U.S. Navy
Molly McCammon, Alaska Ocean Observing System
Walt Meier, National Snow and Ice Data Center
Vera Metcalf, Eskimo Walrus Commission
Son Nghiem, Jet Propulsion Laboratory/Caltech
Jim Overland, NOAA Pacific Marine Environmental Laboratory
Don Perovich, Cold Regions Research and Engineering Laboratory
Andrey Proshutinsky, Woods Hole Oceanographic Institution
Amanda Purcell, National Research Council
Robert Raye, Shell Projects and Technology
Jackie Richter-Menge, Cold Regions Research and Engineering Laboratory
Ignatius Rigor, University of Washington
Mike Terminel, Edison Chouest Offshore
Katie Thomas, National Research Council
Peter Wadhams, University of Cambridge
John Walsh, University of Alaska, Fairbanks
Justin Wettstein, National Center for Atmospheric Research
Jim White, University of Colorado, Boulder
Rebecca Woodgate, University of Washington
Jinlun Zhang, University of Washington

C
Summaries of Workshop Panels and Breakout Discussions

STAKEHOLDERS

The workshop began with a panel discussion on stakeholder needs and key information gaps that could be addressed to help inform the decision-making process. Understanding stakeholder needs is an important step in defining the problem (asking the right questions) and developing solutions. Gary Hufford from the National Oceanic and Atmospheric Administration/ National Weather Service (NOAA/NWS) gave an overview of some of the recent gaps and needs of the Alaska Region NWS program. These include ice coverage and extent needed for crab fishing in the Bering Sea, ice recession time and extent needed for commercial activities, and ice conditions for the safety of subsistence hunters. He noted that there are specific types of information needed to better understand sea ice conditions including seasonal and longer freeze-up and recession times, ice extent, definitive ice type, ice thickness, ice shape, optimized observations, and improved coupled models. Vera Metcalf of the Eskimo Walrus Commission also emphasized that understanding sea ice conditions is critically important for subsistence hunters. They must take a variety of factors into consideration during hunting (which includes towing the mammals onto the ice) and also for their safety. These factors include wind direction and weather, ocean currents, the existence of icebergs, and the thickness of ice.

The US Coast Guard places a great deal of importance on safety in this region, as noted by LCDR Ken Boda. In addition to safety, he emphasized the importance of security (national and economic), stewardship (to protect natural resources and promote science), and operational planning for the future. In terms of seasonal capabilities, the Coast Guard needs to know the timing of sea ice advance and retreat, capabilities for ships, and the footprint of the ice. Understanding what to expect from sea ice on a decadal scale is also important to plan for icebreaking capability and infrastructure in the future. Capt. Michael Terminel from Edison Chouest Offshore pointed out that increasing interest in natural resources is awakening the Arctic region for a number of industry stakeholders. He demonstrated the importance of forecasting weather conditions, and indicated that satellite and radar imagery is a key factor in tracking ice ridges and multiyear ice (this has implications for ship navigation). Like Metcalf, he also noted the critical importance of wind and ocean currents in driving ice conditions.

OBSERVATIONS

During a discussion on key gaps and challenges in observations for future sea ice prediction, Hajo Eicken of the University of Alaska Fairbanks noted a number of gaps in our current understanding of sea ice. These observational challenges include predicting the seasonal decay of ice and heat fluxes over decaying ice. He indicated that when interacting with the modeling community, the decadal timescale is very important, and snow on sea ice is a particularly relevant factor on this scale. He also mentioned that it is important to accurately define the questions that stakeholders would like answered (while also acknowledging that different stakeholders have different needs). He indicated that a path forward may be to define the sea ice services that stakeholders require, translate those needs into specific prognostic variables, and determine the predictive success that is acceptable to the stakeholders.

Walt Meier of the National Snow and Ice Data Center also emphasized the importance of filling gaps in observations, although he pointed out a number of products that are already available for ice extent and concentration, ice thickness, and ice motion. Some current limitations include quantitative error estimates, the harmonization of spatial and temporal scales, melt state and albedo, snow depth, information on ice deformation, in situ and airborne measurements, integrated products, and data access. He also mentioned the importance of continuity and contingency plans in satellite missions (to avoid large gaps in data availability). Ron Lindsay of the University of Washington discussed the importance of in-depth conversations between stakeholders and researchers. This is essential, not only to be sure that we are getting the observations that we actually need, but also to determine where some possible improvements in skill would most help the stakeholders. He noted that, using this information, we can begin to focus on those problems that we can more readily solve and would also be most helpful (instead of using this time on questions that have a limited likelihood of being solved, or on issues that are of limited importance to stakeholders).

Following the panel discussion, members of the breakout groups convened to address additional challenges and strategies associated with sea ice observations. Breakout group rapporteurs mentioned that there is a wide range of needs on many spatial and temporal scales, and that key parameters should be clearly defined depending on stakeholder needs. One group suggested that there are linkages between the need for specific ice parameters and broader scale questions such as: Is there ice? What is it like? Where is it going? This helps to drive observational needs for defining ice extent, ice character, and ice motion. Other issues that were discussed in the breakout groups include quantifying uncertainty, assimilating observations into models, improving bathymetry data, the need for long-term and sustained observations, contingency plans for satellite systems, and better coordination within the modeling, observational, and stakeholder communities.

MODELING

The modeling panel discussion focused on the challenges in modeling and ways to improve interaction with the observational community. Cecilia Bitz of the University of Washington suggested that trying to initialize a model based entirely on observations is a current gap in our understanding. Fully coupled numerical and statistical models will need to be used for prediction, and observations are needed for initial conditions and validation of predictions. She indicated that an important path forward is to communicate the limits of predictability to stakeholders and others, and to use models to determine the most needed measurements. A focus on the coupling between ice, atmosphere, and the ocean was discussed by Elizabeth Hunke from Los Alamos National Laboratory. She indicated that better observations are needed (particularly snow on ice) to continue to make model improvements. She also noted that sea ice predictability is critically dependent on the predictability of the applied forcing and the ice equilibrium state associated with the applied forcing. The strength of feedbacks (including atmosphere and ocean fluxes) needs to be better understood.

Andrey Proshutinsky from the Woods Hole Oceanographic Institution commented on the oceans role in sea ice changes (including three components of influence: atmospheric circulation changes, heat release and ice melt, and sea ice dynamics). He noted that the modeling challenges include reanalysis and reconstruction of sea ice and ocean conditions, implementation of high-resolution models, climate and process studies based on modeling, and systematic model calibration and validation. Possible advances in modeling could address these issues. Examples include increased model resolution, development of a landfast ice model, inclusion of tidal and atmospheric pressure forcing, and improvements of data assimilation methods. Like many of the other panelists, he also mentioned the importance of collaboration with the observational community and stakeholders.

During the breakout group sessions, participants noted that there is a strong stakeholder need for both seasonal predictions (used in planning for fishing, research cruises, industry activities, etc.) and decadal predictions (used in infrastructure planning, national security planning, environmental assessments, endangered species status, etc.). The groups indicated that making predictions on the interannual time frame is particularly difficult. Suggested next steps include work on the question of atmospheric forcing (this can help bridge the gap between the seasonal and decadal timescales), and on the treatment of the ocean in models (models cannot currently resolve vertical stratification and heat fluxes). Other topics that the groups discussed include the transition from first-year ice to multiyear ice, the type of model (statistical, physical) and initializations that should be used for seasonal predictions, coordination opportunities with stakeholders and observationalists, and sensitivity and process studies (including oceanic and atmospheric components) that will improve particular types of forecasts.

CHALLENGES AND OPPORTUNITIES

The workshop concluded with a discussion on the challenges and opportunities in sea ice prediction in the coming decades. Cross-cutting issues and coordination opportunities were discussed by the panelists and participants. Jim Overland from NOAA/ Pacific Marine Environmental Laboratory suggested that researchers should focus on the specific science question of why the ice extent is so low (related to the existence of thin, mobile sea ice). He also noted that models have improved in the past five years in certain aspects, but that significant work remains, particularly in reducing the range of model projections. He points out that, in the "real world" Arctic region, there are important dynamic changes occurring on scales that we cannot yet measure (e.g., large temperature anomalies across small areas). Brendan Kelly of Interagency Arctic Research Policy Committee brought the discussion back to the importance of sustained dialog and in-depth conversations with stakeholders. He noted the significance of making sure that we are asking the right questions and that we are defining the problem appropriately to address stakeholder needs. He challenged the participants to act as ambassadors to their communities in an effort to help others understand the depth and scope of stakeholder needs related to sea ice prediction.

Pablo Clemente-Colón of the National Ice Center commented on the rapidly changing seasonality of Arctic sea ice conditions and the impact that it will have on the type and frequency of measurements being taken. He noted that changes in the amount and location of multiyear ice will directly affect the placement of buoys, for example. To help solve this problem, he suggested that new strategies for in situ observing capabilities will need to be developed, in addition to improvements of currently existing observing systems that could include the integration of new unmanned airborne system and autonomous underwater vehicle technologies. Jean-Claude Gascard of the Universite Pierre et Marie Curie agreed that the main parameters characterizing Arctic sea ice have changed drastically during the past 30 years, and that powerful feedback mechanisms link sea ice with the atmosphere and ocean. He reiterated the significance of the reduction of multiyear ice and noted that a thinner sea ice regime can exhibit less predictability than a thicker regime. He also pointed out the need for process-oriented studies in the atmospheric, sea ice, and oceanic domains, but acknowledged that there are still questions related to what we can predict and how well it can be done. Opportunities for collaboration between observationalists and modelers exist in addressing the issue of trends in variability, with specificity by region. He noted that the data collected over the past 30 years provides unique opportunities for synergies between the observational community and modelers (reanalysis of the past 30 years is important to improve model prediction in the future).

During the final breakout group discussion session, participants noted the need for an ongoing and sustained conversation with stakeholders on what data they want, what they need, and what they can use coupled with a conversation of what forecasters might currently provide, what they might provide soon, and the associated challenges. The groups also noted the importance of changing ice conditions in the Arctic (e.g., the transition from multiyear ice to first-year ice), the need to improve observational capabilities to meet modeling needs, the issue of data continuity, and the opportunity to take advantage of new technologies and collaborations to increase our current understanding. Participants acknowledged that the Arctic is a complex, integrated system including ocean, ice, and atmospheric components and should be treated as an integrated whole.

D
Committee and Staff Biographical Sketches

COMMITTEE

Jacqueline Richter-Menge (Cochair) is a research civil engineer at the Cold Regions Research and Engineering Laboratory (CRREL). Ms. Richter-Menge has focused her research activities on developing a more comprehensive and quantitative understanding of the Arctic sea ice cover, addressing both dynamic and thermodynamic processes. She is a lead investigator in the National Science Foundation Arctic Observing Network program and, with additional support from the National Oceanic and Atmospheric Administration (NOAA) Climate Program Office, directs the activities of a multiagency team establishing a network of autonomous in situ sea ice mass balance observatories. She is a coordinating editor for the Web-based Arctic Report Card for the NOAA Climate Program Office, chairs the U.S. Submarine Arctic Science Program (SCICEX) Science Advisory Committee, and is the sea ice science team lead for the National Aeronautics and Space Administration Operation Ice Bridge Project. In association with her research, Ms. Richter-Menge has gained significant first-hand Arctic experience leading or participating in more than 15 field programs. She actively participates in a wide range of outreach activities, including the coordination of the Adopt-A-Buoy project aimed at middle school science students. Ms. Richter-Menge graduated with a Master of Civil Engineering from the University of Delaware and has been with CRREL since 1981.

John Walsh (Cochair) is a President's Professor of Global Climate Change at the University of Alaska, Fairbanks (UAF). He is also the director of the NOAA/UAF Cooperative Institute for Alaska Research and of the Center for Global Change. His primary research interests are Arctic climate change over the decade-to-century timescale; predictability of climate change in high latitudes, sea ice variations; and extreme weather events in the context of climate change. He was the lead author for the cryosphere chapter of the *Arctic Climate Impact Assessment* (2005) and a lead author for the Polar Regions chapter of the IPCC's Fourth Assessment Report (2007). He is a coordinating lead author for the 2013 National Assessment Report being produced by the U.S. Global Change Research Program. Prior to his position at the University of Alaska, Walsh spent 30 years on the faculty of the University of Illinois at Urbana. He is the coauthor of an undergraduate textbook on severe and hazardous weather. He earned his Ph.D. in meteorology from the Massachusetts Institute of Technology in 1974 and his B.A. from Dartmouth College in 1970.

Lawson Brigham is Distinguished Professor of Geography & Arctic Policy at the University of Alaska Fairbanks, and a senior fellow at the Institute of the North in Anchorage. During 2005-2009 he was chair of the Arctic Council's Arctic Marine Shipping Assessment and vice chair of the council's working group on Protection of the Arctic Marine Environment. Dr. Brigham was a career U.S. Coast Guard officer, serving from 1970 to 1995 and retiring with the rank of Captain. He served at sea in command of four Coast Guard cutters including a patrol boat, Great Lakes icebreaker, offshore law enforcement cutter, and the polar icebreaker *Polar Sea* sailing in Alaskan, Arctic, and Antarctic waters; he also served as chief of strategic planning in Washington, D.C. Dr. Brigham has been a research fellow at Woods Hole Oceanographic Institution, a faculty member of the U.S. Coast Guard Academy and the Naval Postgraduate School, and deputy director of the U.S. Arctic Research Commission. He is a graduate of the U.S. Coast Guard Academy (B.S.), a distinguished graduate of the U.S. Naval War College, and holds graduate degrees from Rensselaer Polytechnic Institute (M.S.) and the University of Cambridge (M.Phil. and Ph.D.). His research interests include Arctic marine transportation, remote sensing of sea ice, Arctic climate change, and polar marine policy.

Jennifer A. Francis is a research professor at the Institute of Marine and Coastal Sciences and the Graduate Program in Atmospheric Sciences at Rutgers University. She studies the Arctic climate system, causes for rapid change, and linkages between the Arctic and the global climate system. Her work is funded primarily by the National Science Foundation. She has served on several national committees in the National Science Foundation, the American Meteorological Society, and the science steering committee for the Study of Arctic Environmental Change (SEARCH). Dr. Francis received her Ph.D. in atmospheric sciences from the University of Washington in 1994. Dr. Francis is currently a member of the Polar Research Board.

Marika Holland is a an ice specialist in the Oceanography section of the Climate and Global Dynamics division at the National Center for Atmospheric Research (NCAR). She received her Ph.D. in 1997 from the Program in Atmosphere and Ocean Sciences at the University of Colorado in the area of sea ice modeling for climate applications. Her training continued with a postdoctoral fellowship at the University of Victoria in British Columbia, studying the influence of sea ice variability and change on the global ocean circulation and climate. In 1999, Dr. Holland moved to NCAR in Boulder, Colorado, as a postdoctoral fellow and joined the scientific staff in 2000. Her research interests include polar climate variability and future change, including the role of ice-ocean-atmosphere interactions and feedbacks. She has extensive experience using coupled climate models to study these issues and has been active in the development of improved sea ice models for climate simulations. She is currently serving as chief scientist for the Community Earth System Modeling Project.

Son V. Nghiem is the Science Applications Development lead of the Radar Science and Engineering Section, and the Hydrology Discipline program manager of the Hydrology Office in the Earth Science and Technology Directorate at the Jet Propulsion Laboratory of the California Institute of Technology. His research encompasses active and passive remote sensing, advanced satellite radars and radiometers, electromagnetic scattering and emission, and earth sciences and applications. He has published 70 peer-reviewed articles and over 230 conference articles. He received the 1999 Lew Allen Award for Excellence in recognition of his pioneering research in the areas of polarimetric scatterometry for earth science remote sensing and contributions to future advanced satellite instrument concepts; the 2006 NASA Exceptional Achievement Medal for developing scientific applications of scatterometry in land, ice, and snow processes;

the 2008 NASA Exceptional Scientific Achievement Medal for his contributions to understanding the melt state of Greenland and Antarctica ice sheets, its significance in earth science missions, and its implications in climate change; and the 2010 NASA Exceptional Technology Achievement Medal for his contributions in developing a new technology using NASA satellite scatterometer data to measure high-resolution global wind for offshore wind energy development. His research results were reported worldwide by major news networks and many radio stations. Dr. Nghiem received his Ph.D. from the Massachusetts Institute of Technology in 1991.

Robert Raye is the Ice and Metocean Project lead for Shell Projects and Technology in the U.S. Arctic. In this role, Mr. Raye is responsible for providing support to field activities and design engineering to ensure safe and efficient operations. He has a key role in delivery of Shell's Arctic physical science program, which includes collection of field measurements, characterization and research studies, and collaborative programs with industry partners, academia, and governmental agencies. Mr. Raye has established a field observation program in Alaska that includes a network of instrumented buoys, coastal meteorological stations, and vessel-based observers that report near-real-time data used to validate models and forecasts. Recently, he has been instrumental in developing collaborative agreements with NOAA offices to share data and resources, with the goal of improving overall weather and ice forecasting in Alaska and improving hurricane intensity forecasting in the Gulf of Mexico. He serves on the Data Management and Communications Committee in the Gulf Coast Ocean Observing System, where he supports initiatives promoting data interoperability, metadata standards, and Web services for data products and has applied these concepts in Shell internal data management and dissemination systems. Mr. Raye is Shell's subject matter expert for oceanographic surveys and is skilled in environmental instrumentation, data analyses, and data management. Mr. Raye holds a Master of Science degree in ocean engineering from Florida Atlantic University.

Rebecca Woodgate[1] is a principal oceanographer and associate professor at the Applied Physics Laboratory and the School of Oceanography at the University of Washington. She is a physical oceanographer, specializing in polar research, with special focus on the circulation of the Arctic Ocean, interactions between sea ice and the ocean, and the role of the polar oceans in climate. Her research concentrates on the collection and analysis of in situ oceanographic data. She has worked for many years in the deployment and recovery of moored oceanographic instrumentation in ice-covered waters, and the analysis of both mooring and hydrographic data. She is involved in undergraduate teaching and graduate education. She has worked on British, German, Norwegian, and American research vessels and led expeditions to the Bering Strait and the Arctic Ocean. Her first degree is in physics from the University of Cambridge and her Ph.D. (University of Oxford) is in data assimilation in ocean models. Her postdoctoral work was done at the Alfred Wegener Institute in Germany. Dr. Woodgate's research goal is to understand the physical processes in both Arctic and Antarctic regions, and to use her background to bridge the gap between theory, modeling, and real observations of the oceans.

[1] Member through June 2012

NATIONAL RESEARCH COUNCIL STAFF

Ms. Katie Thomas is an associate program officer for the Board on Atmospheric Sciences and Climate (BASC). She received her B.S. from the University of Michigan in 2004 and her M.S. in environmental science and policy from Johns Hopkins University in 2009. Since joining the National Research Council in 2006, she has worked on studies related to urban meteorology, climate modeling, weather radar, and advancing climate science.

Ms. Lauren Brown is a research associate with the Polar Research Board and the Board on Atmospheric Sciences and Climate at the National Academies, where she has been involved in a number of National Research Council studies such as *America's Climate Choices*, *Lessons and Legacies of International Polar Year 2007-2008*, and *Future Science Opportunities in Antarctica and the Southern Ocean*. She holds an M.S. in marine studies with a focus on physical ocean science and engineering and a B.A. in physics and astronomy from the University of Delaware. She is especially interested in high-latitude environmental policy issues and the role of polar regions in global climate change.

Ms. Amanda Purcell is a research associate with the Board on Atmospheric Sciences and Climate (BASC). She began working with BASC as a program assistant in 2008 and has since worked on various projects including America's Climate Choices, Frontiers in Understanding Climate Change and Polar Ecosystems, and Future Science Opportunities in Antarctica and the Southern Ocean. Amanda received her bachelor's degree in physics and mathematics from American University in 2008. She is also currently pursuing a master's in mathematics from American University, anticipated in 2013.

Dr. Alexandra Jahn is a project scientist at the National Center for Atmospheric Research (NCAR). Her research interests are in Arctic sea ice and freshwater dynamics, climate modeling, ocean tracers, and paleoclimate. Alexandra received her Ph.D. in atmospheric and oceanic sciences from McGill University in 2010, for her research on Arctic Ocean freshwater dynamics. After a 2-year postdoctoral appointment in the Advanced Study Program at NCAR, Alexandra was a Christine Mirzayan Science Policy Fellow with the National Research Council's Polar Research Board in early 2012, before returning to NCAR for her current appointment.

Ms. Elizabeth Finkelman is a senior program assistant for the Board on Atmospheric Sciences and Climate (BASC). She received her Bachelor of Arts and Science degree from McGill University in 2010, concentrating in molecular biology and political science. Since joining the National Research Council in March 2011, she has participated in board-related projects and studies concerning climate change, urban meteorology, climate modeling, and urban forestry.